MISSION ACCOMPLISHED
with
Engineering Design Graphics

A Workbook for Future Engineers
Featuring
Engineering Sketching, Computer Aided Design
and Solid Models

Gerald E. Vinson, Ed.D.

Program Coordinator, Engineering Design Graphics
Texas A&M University
College Station, Texas, 77843-3138

KENDALL/HUNT PUBLISHING COMPANY
4050 Westmark Drive Dubuque, Iowa 52002

Other workbooks in this Engineering Graphics series by Gerald E. Vinson:

- From Sketch Pad to AutoCAD with Solid Models, 1998
- Creative Engineering Graphics, 1999
- Engineering Design Graphics for a Changing World, 2000

All published by Kendall/Hunt Publishing Company, Dubuque, Iowa

Disclaimer: The problems in this workbook are intended for drawing practice only. The author or publisher assumes no liability for the manufacturing of any objects or assemblies shown.

Cover artwork of the shuttle Endeavor courtesy of the NASA Johnson Spaceflight Center.

ISBN 0-7872-9406-3

Printed in the United States of America
10 9 8 7 6 5 4 3

TO THE STUDENT

This workbook is designed to help you communicate more effectively as a beginning engineer or designer through the use of engineering design graphics. The problems include practice in graphics theory, use of "ANSI" standards and open-ended design concepts to stimulate creativity and provide realistic practice in problem solving. Each problem sheet will "ramp" your abilities up from simple to more complex solutions. The solutions are first sketched freehand (or with minimal drafting aids) and then converted to computer drawings. The computer drawings are created with AutoCAD® software and are best constructed on similar software. The template files provide freestanding drawings that can be printed to accurate scale sizes on most printers or pen plotters using plain paper. Optimum plots are produced on laser printers.

COURSE ORGANIZATION

This course is designed to teach basic drafting conventions and sketching skills used for the construction of computer-aided drafting and design "CAD" drawings. Instructors may assign all or only portions of each problem sheet in order to optimize the learning experience. Occasionally, students will be required to work in design teams or pairs. Sketches are created on sheets with .10" grid patterns to make layouts more accurate and easier to draw. In most instances, both the layout sketch and the final computer prints are turned in together. Occasionally, only one or the other will be assigned. Many problems are also suitable for drawing as solid models if designated by your instructor. The course grade will be determined using the percentages similar to the sample below.

A SAMPLE GRADE CALCULATION

MAJOR TOPICS	% OF GRADE	YOUR AVERAGE	POINTS EARNED
Daily Exercises	30%	95	28.5
Quiz Average	30%	86	25.8
Design Project	20%	90	18.0
Design & WD1	10%	92	9.2
Design & WD2	10%	95	9.5
Maximum	100%		91.0

PROBLEM REFERENCE GUIDE

TOPIC & PROBLEM NUMBERS		G.E. Vinson[1]	J. H. Earle[2]	F. Giesecke[3]	G.R. Bertoline[4]
		CHAPTER	CHAPTER	CHAPTER	CHAPTER
Lettering Practice	1-4, 25, 39	1	11	4	4
Basic Sketching	5-6, 12	2	13	6	4
Use of Scales	7,8,55	4	10	3	3
Basic Multi-view Sketching	5-13	3	13	7	8
Orthographic Views	13-22	3	14	7	4
Section Views	44-50	6	16	8	14
Crosshatch Patterns	40-43	6	16	8	14
Threads and Fasteners	52-54	5	17	14	15
Auxiliary Views	73-75	6	15	9	11
Dimensioning	56-63	4	19	12	15
Tolerance Specifications	64-66	7	20	13	15
Geometric Tolerances	67-69	7	20	13	16
Design / Working Drawings	70-72	8 & 9	22	15	19
Oblique Pictorials	26-32	2	24	18	9
Isometric Pictorials	33-38	2	24	17	9
Design Modification	23, 24, 32	8	9	15	19
Standard Hole Treatments	44	4	20	11, 12	18
Standard Features	51	4	14	12	18
Solid Models in CAD	74-85	10	30	--	7
Design Team Process	Dp 1-15	9	4, 9	1	2

[1] G.E. Vinson, ESSENTIALS OF ENGINEERING DESIGN GRAPHICS, 2ND Ed., Kendall/Hunt Publishing Company, 2003.

[2] J. H. Earle, GRAPHICS FOR ENGINEERS, 5TH, Ed., Prentice Hall, Inc., 2000.

[3] Fredrick E. Giesecke, et al, TECHNICAL DRAWING, 10TH Ed., Prentice Hall, Inc., 1997.

[4] G.R. Bertoline, et al, TECHNICAL COMMUNICATIONS, 2nd Ed., Irwin, 1997

COMMON GRADING ABBREVIATIONS

One of the most efficient methods to improve your drawing skill is to carefully review graded drawings as they are returned. Recognizing mistakes before they become bad habits is essential to continual improvement and quality work.

To speed up the turn-a-round time on corrected drawings, your instructor may use abbreviations such as:

1. **LQ**......... Line Quality
2. **AH**........ Arrowheads
3. **W**.......... Weak Lines
4. **H**........... Heavy Line
5. **SP**.......... Spelling Error
6. **E**............ Erase
7. **O**........... Circle Over Bad Junctions
8. **CONST**...... Poor Construction
9. **ACCY**........ Improve Accuracy
10. **LTR**............ Lettering
11. **SPC**............ Space Needed
12. **LTYP**......... Linetype
13. **GL**.............. Guidelines Needed
14. **CWD**.......... Crowded Views or Dimensions

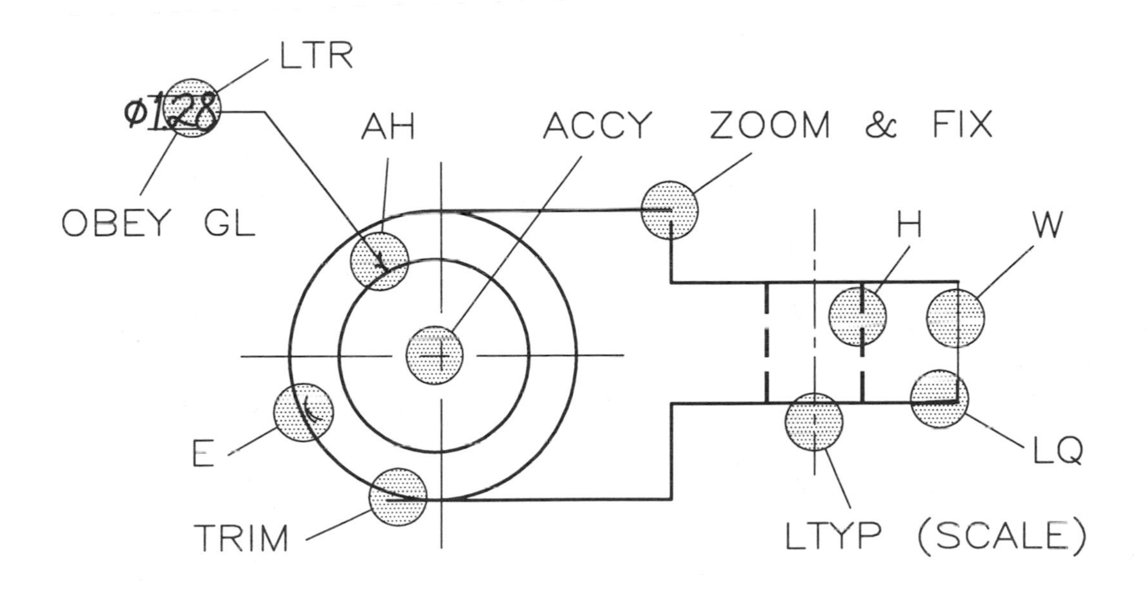

Student's Grade Summary Sheet

PROBLEM NUMBER	DUE DATE	GRADE EARNED

PROBLEM NUMBER	DUE DATE	GRADE EARNED
PROBLEM AVG:		

TEST SCORES
1.
2.
3.
4.
5.
6.
7.
8.
9.
TOTAL:
AVG:

DESIGN PROJ.
I
II
III
IV
V
VI
AVG:

DESIGN DWG.
I
II
AVG:

Your final grade is determined by the percentages outlined in the course syllabus.

1 LETTERING PRACTICE WITH VERTICAL GOTHIC

INSTRUCTIONS: USE AN F, H, OR HE PENCIL TO PRINT AT LEAST SIX REPETITIONS OF EACH LETTER SHOWN. USE GUIDELINES AND ALL CAPS.

A A A A
A A A

A B C D E

F G H I J

K L M N O

P Q R S T

U V W X Y

Z & %

USE ONLY CAPITAL LETTERS.

1 2 3 4 5

6 7 8 9 10

THE LETTERING HEIGHT FOR METRIC DRAWINGS IS 3mm (.125 INCHES).

FILE:	DRAWN BY:	TEAM NO.	IMPROVEMENT NEEDED: ☐ 1. LINE QUALITY ☐ 2. LETTERING ☐ 3. CONSTRUCTION ☐ 4. ACCURACY	1
GRADE:	SECTION: DATE DUE:	SCALE:		gev-e

0 1 2 3 4 5 6 7 8 9

10 SCALE — FULL SIZE INCHES

0 1 2 3 4 5 6 7 8 9 10 11 12 13 14 15 16 17 18

20 SCALE — 1 INCH EQUALS 20 UNITS.

0 2 4 6 8 10 12 14 16 18 20 22 24 26

30 SCALE — 1 INCH EQUALS 30 UNITS.

0 2 4 6 8 10 12 14 16 18 20 22 24 26 28 30 32 34 36

40 SCALE — 1 INCH EQUALS 40 UNITS.

0 2 4 6 8 10 12 14 16 18 20 22 24 26 28 30 32 34 36 38 40 42 44

50 SCALE — 1 INCH EQUALS 50 UNITS.

0 2 4 6 8 10 12 14 16 18 20 22 24 26 28 30 32 34 36 38 40 42 44 46 48 50 52 54

60 SCALE — 1 INCH EQUALS 60 UNITS.

COMMON ENGINEERS' SCALES

USE THESE SCALES TO ESTABLISH DIMENSIONS ON DRAWING EXERCISES.

1C

gev–e

(2) LETTERING PRACTICE WITH INCLINED GOTHIC

INSTRUCTIONS: USE AN F, H, OR HB PENCIL TO PRINT AT LEAST SIX REPETITIONS OF EACH LETTER SHOWN. USE GUIDELINES AND ALL CAPS.

A A A A A A A

A B C D E

F G H I J

K L M N O

P Q R S T

U V W X Y

Z & %

USE ONLY CAPITAL LETTERS.

1 2 3 4 5

6 7 8 9 10

THE STANDARD ANSI ANGLE FOR INCLINED LETTERING IS 68° (DEGREES).

FILE:	DRAWN BY:	TEAM NO.	IMPROVEMENT NEEDED:	2
GRADE:	SECTION: DATE DUE:	SCALE:	☐ 1. LINE QUALITY ☐ 2. LETTERING ☐ 3. CONSTRUCTION ☐ 4. ACCURACY	gev–e

0 1 2 3 4 5 6 7 8 9 10 11 12 13 14 15 16 17 18 19 20 21 22 23
1:1 SCALE FULL SIZE MILLIMETERS

0 5 10 15 20 25 30 35 40 45
1:2 SCALE METRIC SCALE

0 10 20 30 40 50 60
1:3 SCALE METRIC SCALE

0 10 20 30 40 50 60 70 80 90
1:4 SCALE METRIC SCALE

0 10 20 30 40 50 60 70 80 90 100 110
1:5 SCALE METRIC SCALE

0 10 20 30 40 50 60 70 80 90 100 120 130 130
1:6 SCALE METRIC SCALE

COMMON METRIC SCALES

USE THESE SCALES TO ESTABLISH DIMENSIONS ON DRAWING EXERCISES.

2c
gev–e

(3) LETTERING PRACTICE FOR SENTENCES.

INSTRUCTIONS: COPY EACH SENTENCE ONCE OR TWICE AS ASSIGNED BY YOUR TEACHER. BE SURE TO USE PROPER SPACING BETWEEN LETTERS AND WORDS.

1. EXCELLENT LETTERING QUALITY IS POSSIBLE WHEN ONE IS DELIBERATE.

2. DASHES OF HIDDEN LINES ARE APPROXIMATELY .125 INCHES LONG.

3. ENGINEERING GRAPHICS PROVIDE SPECIFICATIONS FROM ENGINEERS.

4. THE ABBREVIATION FOR DIAMETER IS "DIA" OR THE SYMBOL "⌀".

5. ALWAYS USE CAPITAL LETTERS AND GUIDELINES WHEN YOU PRINT.

6. THERE ARE ONLY SIX PRINCIPAL VIEWS IN ORTHOGRAPHIC DRAWINGS.

7. ENGINEERS REQUIRE CLEAR, CONCISE, AND ACCURATE COMMUNICATION.

FILE:	DRAWN BY:	TEAM NO.	IMPROVEMENT NEEDED: ☐ 1. LINE QUALITY ☐ 2. LETTERING ☐ 3. CONSTRUCTION ☐ 4. ACCURACY	3
GRADE:	SECTION: DATE DUE:	SCALE:		gev–e

(3c) TEXT BY COMPUTER

INSTRUCTIONS: TYPE EACH SENTENCE ONCE OR TWICE AS ASSIGNED BY YOUR TEACHER.
USE A DIFFERENT FONT STYLE FOR EACH SENTENCE.

1. EXCELLENT LETTERING QUALITY IS POSSIBLE WHEN ONE IS DELIBERATE.

2. DASHES OF HIDDEN LINES ARE APPROXIMATELY .125 INCHES LONG.

3. ENGINEERING GRAPHICS PROVIDE SPECIFICATIONS FROM ENGINEERS.

4. THE ABBREVIATION FOR DIAMETER IS "DIA" OR THE SYMBOL "ø".

5. ALWAYS USE CAPITAL LETTERS AND GUIDELINES WHEN YOU PRINT.

6. THERE ARE ONLY SIX PRINCIPAL VIEWS IN ORTHOGRAPHIC DRAWINGS.

7. ENGINEERS REQUIRE CLEAR, CONCISE, AND ACCURATE COMMUNICATION.

FILE: 00	DRAWN BY: LAST, FIRST TEAM NO. 00	IMPROVEMENT NEEDED:	3c
GRADE:	SECTION: 500 DATE DUE: 00/00/00 SCALE: FULL	□ 1. LINE QUALITY □ 2. LETTERING □ 3. CONSTRUCTION □ 4. ACCURACY	gev—e

(4) LETTERING PRACTICE FOR SENTENCES.

INSTRUCTIONS: COPY EACH SENTENCE ONCE OR TWICE AS ASSIGNED BY YOUR TEACHER. BE SURE TO USE PROPER SPACING BETWEEN LETTERS AND WORDS.

1. PROPER LEAD SELECTION AND PRESSURE INSURE GOOD LINE QUALITY.
2. CENTERLINES ARE USED TO LOCATE CENTERS OF CIRCULAR FEATURES.
3. ORTHOGRAPHIC VIEWS INCLUDE: FNT & REAR, TOP & BOT, LEFT & RT.
4. THE ABBREVIATION FOR RADIUS IS "R" AND FOR ROUND IS "RND".
5. SOME OBJECTS CAN BE FULLY DESCRIBED IN LESS THAN THREE VIEWS.
6. PRINT YOUR NAME, FILE, AND SECTION NUMBER ON EACH ASSIGNMENT.
7. QUICK ENGINEERING SKETCHES COMMUNICATE THE DESIGN INTENT.

FILE:	DRAWN BY:	TEAM NO.	IMPROVEMENT NEEDED: ☐ 1. LINE QUALITY ☐ 2. LETTERING ☐ 3. CONSTRUCTION ☐ 4. ACCURACY	4 gev–e
GRADE:	SECTION: DATE DUE:	SCALE:		

(4c) TEXT BY COMPUTER

INSTRUCTIONS: TYPE EACH SENTENCE ONCE OR TWICE AS ASSIGNED BY YOUR TEACHER. USE A DIFFERENT FONT STYLE FOR EACH SENTENCE.

1. PROPER LEAD SELECTION AND PRESSURE INSURE GOOD LINE QUALITY.
2. CENTERLINES ARE USED TO LOCATE CENTERS OF CIRCULAR FEATURES.
3. ORTHOGRAPHIC VIEWS INCLUDE: FNT & REAR, TOP & BOT, LEFT & RT.
4. THE ABBREVIATION FOR RADIUS IS "R" AND FOR ROUND IS "RND".
5. SOME OBJECTS CAN BE FULLY DESCRIBED IN LESS THAN THREE VIEWS.
6. PRINT YOUR NAME, FILE, AND SECTION NUMBER ON EACH ASSIGNMENT.
7. QUICK ENGINEERING SKETCHES COMMUNICATE THE DESIGN INTENT.

FILE: 00	DRAWN BY: LAST, FIRST	TEAM NO. 00	IMPROVEMENT NEEDED:	4c
GRADE:	SECTION: 500 DATE DUE: 00/00/00	SCALE: FULL	☐ 1. LINE QUALITY ☐ 2. LETTERING ☐ 3. CONSTRUCTION ☐ 4. ACCURACY	gev-e

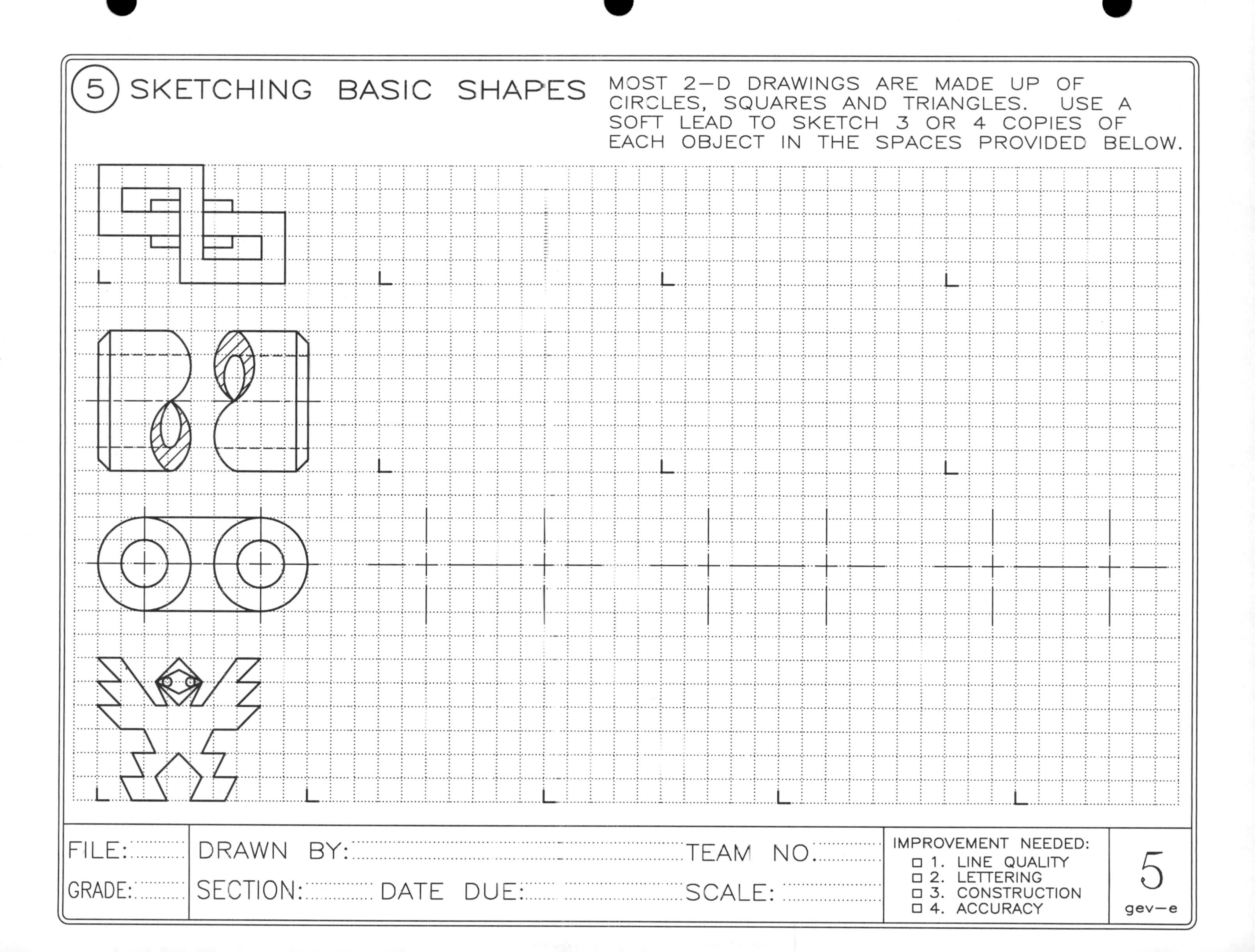
5 SKETCHING BASIC SHAPES
MOST 2–D DRAWINGS ARE MADE UP OF CIRCLES, SQUARES AND TRIANGLES. USE A SOFT LEAD TO SKETCH 3 OR 4 COPIES OF EACH OBJECT IN THE SPACES PROVIDED BELOW.
FILE:
GRADE:
DRAWN BY:
TEAM NO.
SECTION:
DATE DUE:
SCALE:
IMPROVEMENT NEEDED:
☐ 1. LINE QUALITY
☐ 2. LETTERING
☐ 3. CONSTRUCTION
☐ 4. ACCURACY
5
gev–e

(5c) BASIC COMPUTER COMMANDS

USE AutoCAD COMMANDS: ARC, CIRCLE, LINE, UNDO, AND ERASE (ALIAS A-CLUE) TO DRAW THE OBJECTS SKETCHED ON THE OTHER SIDE OF THIS PAGE. DRAW A CHILD'S WAGON AND A TOY CAR IN THE SPACES PROVIDED.

WAGON

TOY CAR

FILE: 00	DRAWN BY: LAST, FIRST	TEAM NO. 00	IMPROVEMENT NEEDED:	5c
GRADE:	SECTION: 500 DATE DUE: 00/00/00	SCALE: FULL	☐ 1. LINE QUALITY ☐ 2. LETTERING ☐ 3. CONSTRUCTION ☐ 4. ACCURACY	gev-e

(6) BASIC SKETCHING PRACTICE: *ARROW TACKER/STAPLER*

SKETCH AN ORTHOGRAPHIC VIEW OF THE ARROW STAPLER SHOWN BELOW.
NOTE: MANY OF THE HIDDEN LINES HAVE BEEN OMITTED FOR CLARITY.

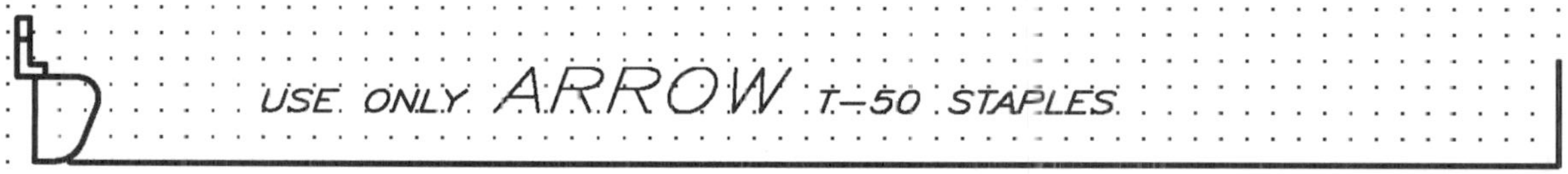

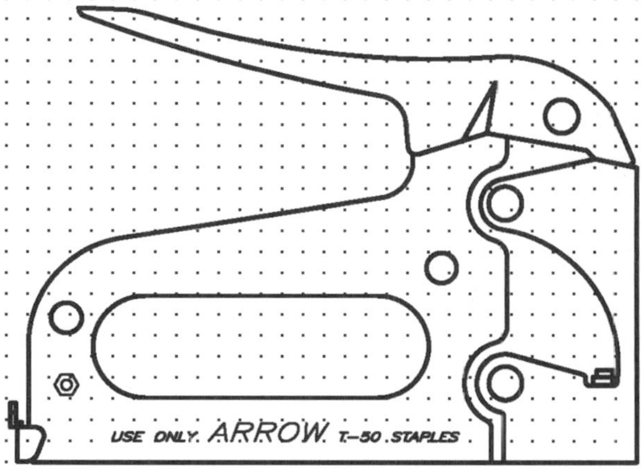

USE ONLY ARROW T-50 STAPLES

USE ONLY ARROW T-50 STAPLES

FILE:	DRAWN BY:	TEAM NO.	IMPROVEMENT NEEDED: ☐ 1. LINE QUALITY ☐ 2. LETTERING ☐ 3. CONSTRUCTION ☐ 4. ACCURACY	6
GRADE:	SECTION: DATE DUE:	SCALE:		gev-e

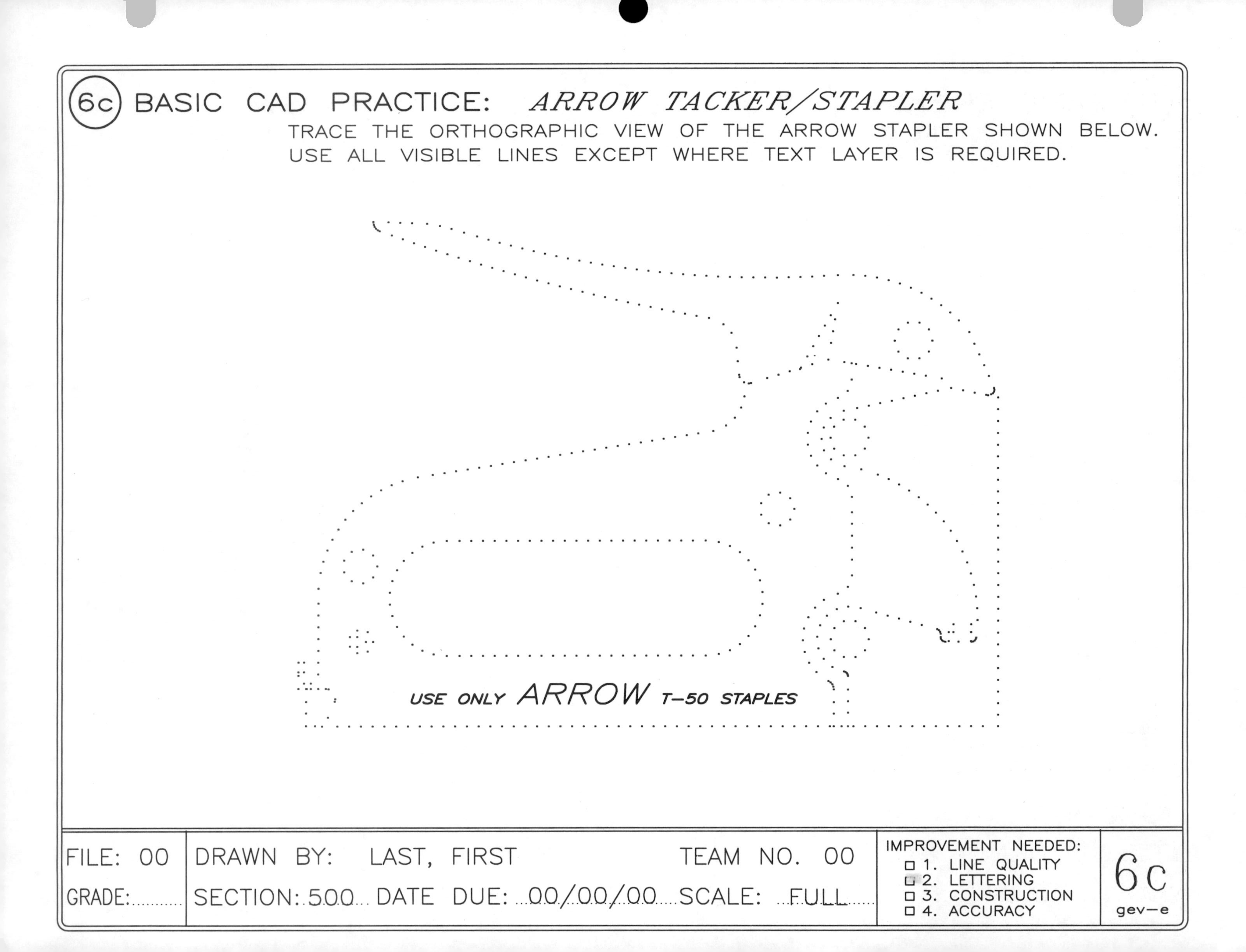
6c
BASIC CAD PRACTICE: ARROW TACKER/STAPLER
TRACE THE ORTHOGRAPHIC VIEW OF THE ARROW STAPLER SHOWN BELOW.
USE ALL VISIBLE LINES EXCEPT WHERE TEXT LAYER IS REQUIRED.
USE ONLY ARROW T-50 STAPLES
FILE: 00
GRADE:
DRAWN BY: LAST, FIRST
TEAM NO. 00
SECTION: 500
DATE DUE: 00/00/00
SCALE: FULL
IMPROVEMENT NEEDED:
1. LINE QUALITY
2. LETTERING
3. CONSTRUCTION
4. ACCURACY
6c
gev-e

(8) SCALE EXERCISE

GIVEN THAT THE TRUE SIZE OF THE SMOOTHING PLANE BASE IS 150mm, DETERMINE THE SCALE OF THIS DRAWING AND NEATLY PRINT IN THE 10 MISSING DIMENSIONS. THE PLANE IS A STANLEY NO. 02 FROM AROUND 1885.

SCALE: ....

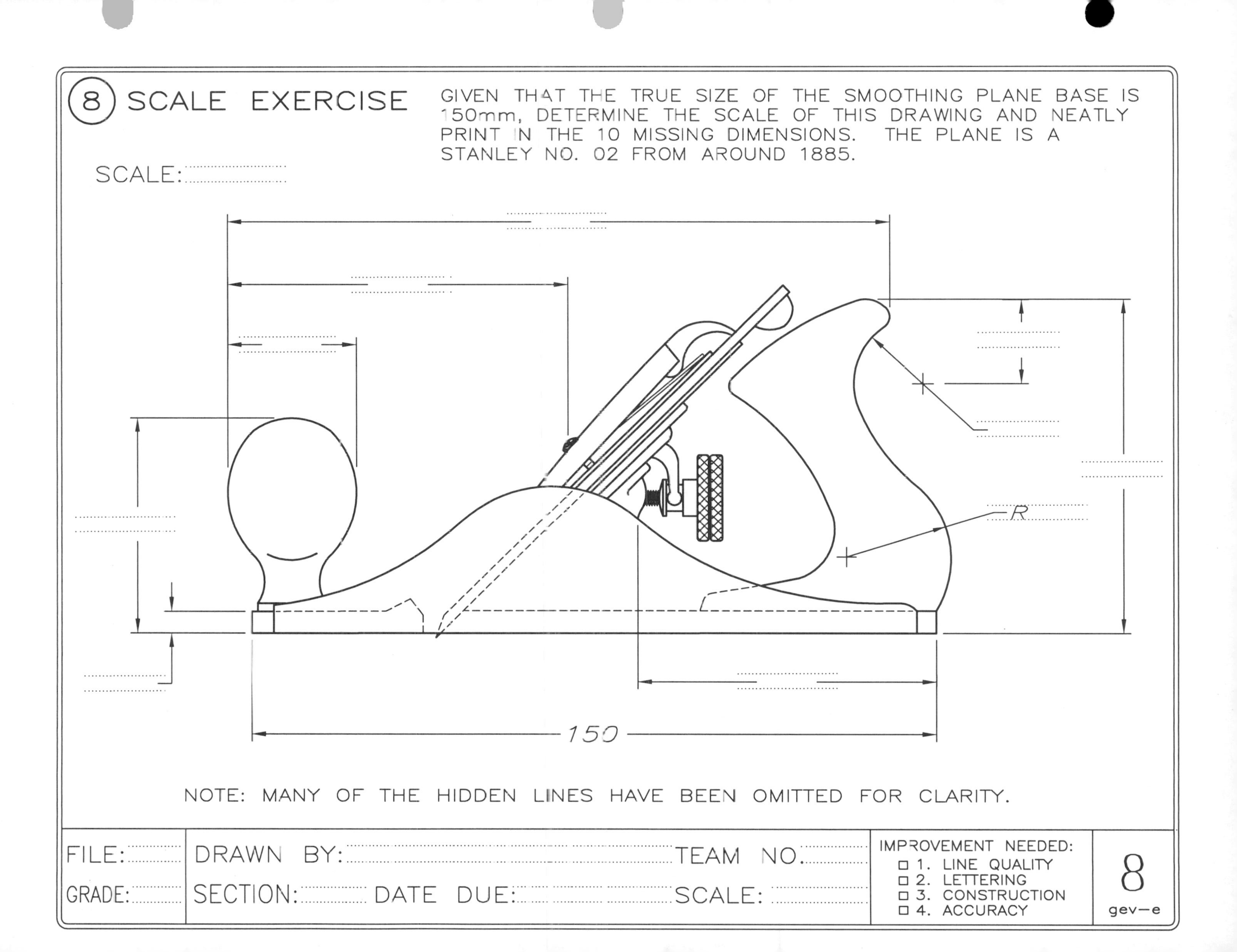

NOTE: MANY OF THE HIDDEN LINES HAVE BEEN OMITTED FOR CLARITY.

FILE:	DRAWN BY:	TEAM NO.	IMPROVEMENT NEEDED: ☐ 1. LINE QUALITY ☐ 2. LETTERING ☐ 3. CONSTRUCTION ☐ 4. ACCURACY	8 gev–e
GRADE:	SECTION: DATE DUE:	SCALE:		

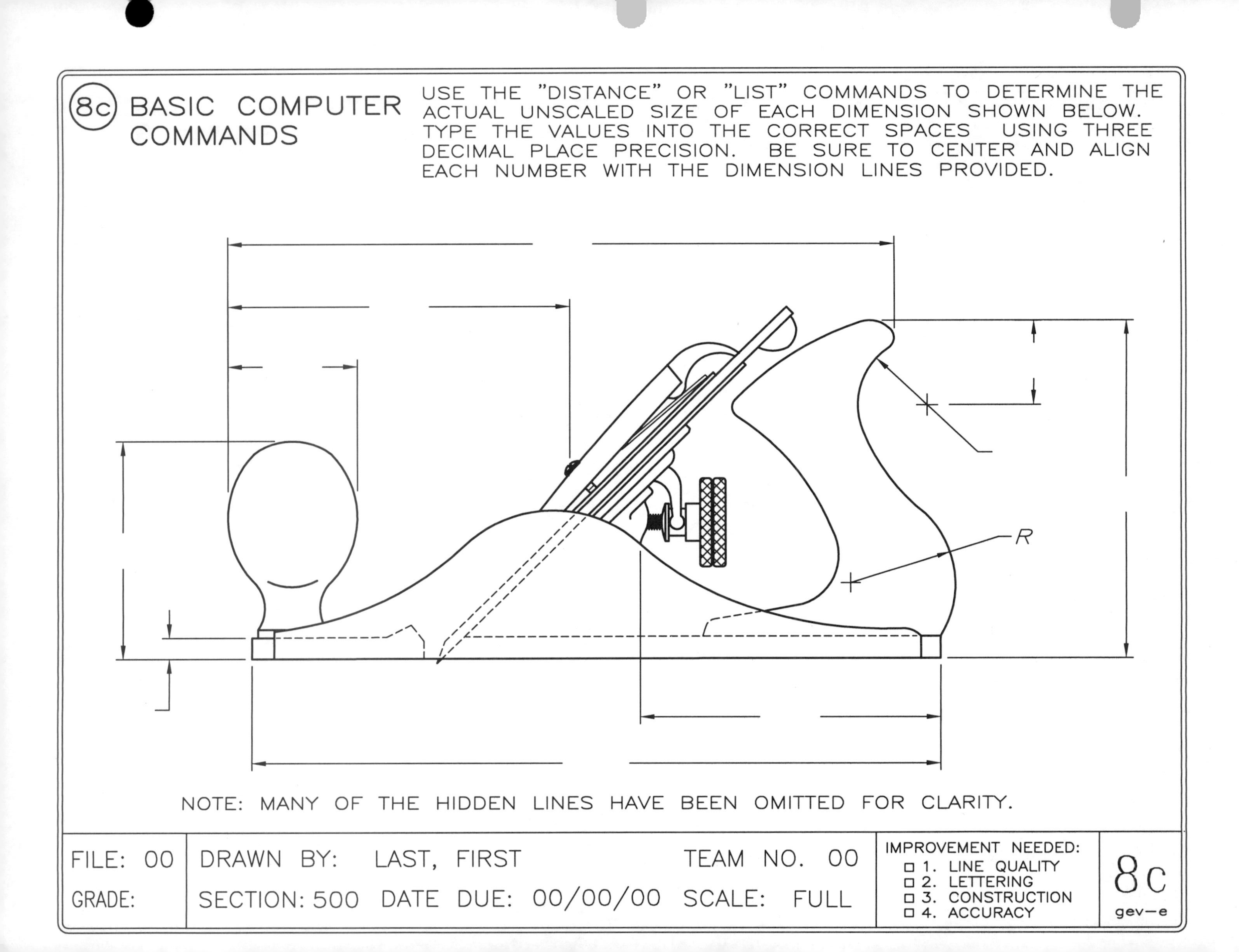
8c
BASIC COMPUTER COMMANDS
USE THE "DISTANCE" OR "LIST" COMMANDS TO DETERMINE THE ACTUAL UNSCALED SIZE OF EACH DIMENSION SHOWN BELOW. TYPE THE VALUES INTO THE CORRECT SPACES USING THREE DECIMAL PLACE PRECISION. BE SURE TO CENTER AND ALIGN EACH NUMBER WITH THE DIMENSION LINES PROVIDED.
R
NOTE: MANY OF THE HIDDEN LINES HAVE BEEN OMITTED FOR CLARITY.
FILE: 00
GRADE:
DRAWN BY: LAST, FIRST
TEAM NO. 00
SECTION: 500
DATE DUE: 00/00/00
SCALE: FULL
IMPROVEMENT NEEDED:
□ 1. LINE QUALITY
□ 2. LETTERING
□ 3. CONSTRUCTION
□ 4. ACCURACY
8c
gev–e

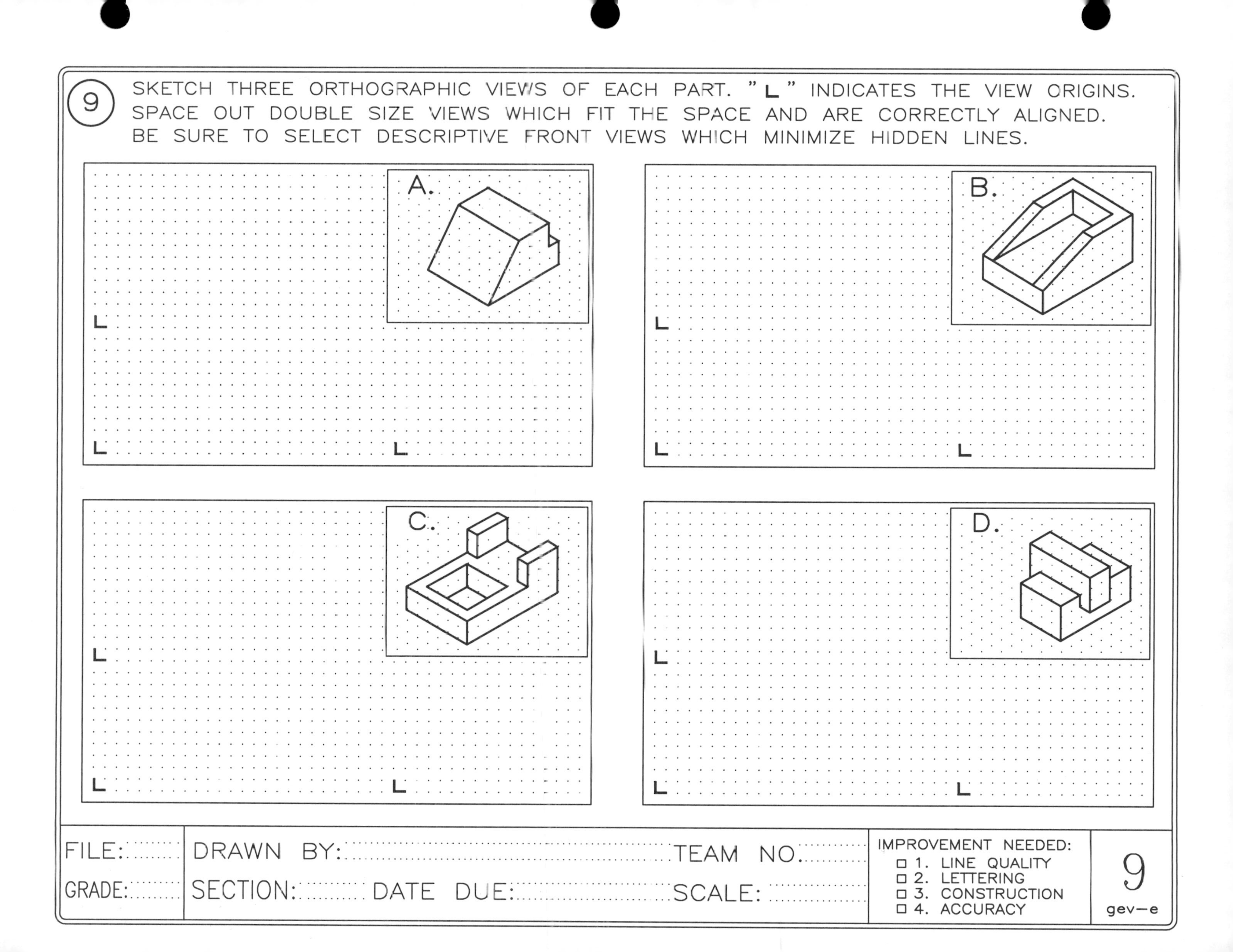
9
SKETCH THREE ORTHOGRAPHIC VIEWS OF EACH PART. "L" INDICATES THE VIEW ORIGINS.
SPACE OUT DOUBLE SIZE VIEWS WHICH FIT THE SPACE AND ARE CORRECTLY ALIGNED.
BE SURE TO SELECT DESCRIPTIVE FRONT VIEWS WHICH MINIMIZE HIDDEN LINES.
A.
B.
C.
D.
FILE:
GRADE:
DRAWN BY:
SECTION:
DATE DUE:
TEAM NO.
SCALE:
IMPROVEMENT NEEDED:
1. LINE QUALITY
2. LETTERING
3. CONSTRUCTION
4. ACCURACY
9
gev-e

9c MULTI—VIEW DRAWINGS

REFER TO YOUR SOLUTIONS ON THE OTHER SIDE OF THIS PAGE TO MAKE THREE VIEW ORTHOGRAPHIC DRAWINGS OF THE OBJECTS BELOW.

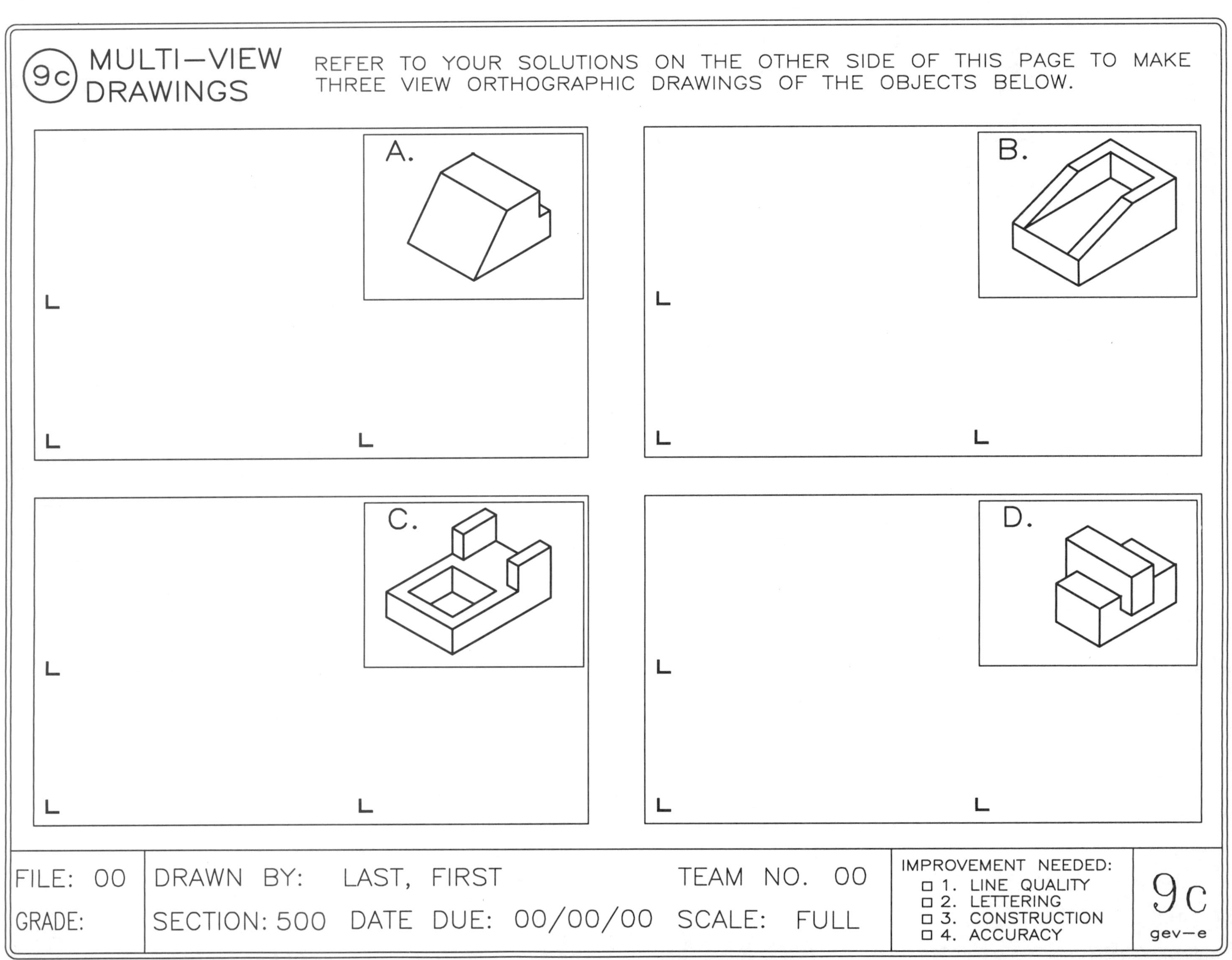

FILE: 00 | DRAWN BY: LAST, FIRST | TEAM NO. 00

GRADE: | SECTION: 500 | DATE DUE: 00/00/00 | SCALE: FULL

IMPROVEMENT NEEDED:
- ☐ 1. LINE QUALITY
- ☐ 2. LETTERING
- ☐ 3. CONSTRUCTION
- ☐ 4. ACCURACY

9c

gev—e

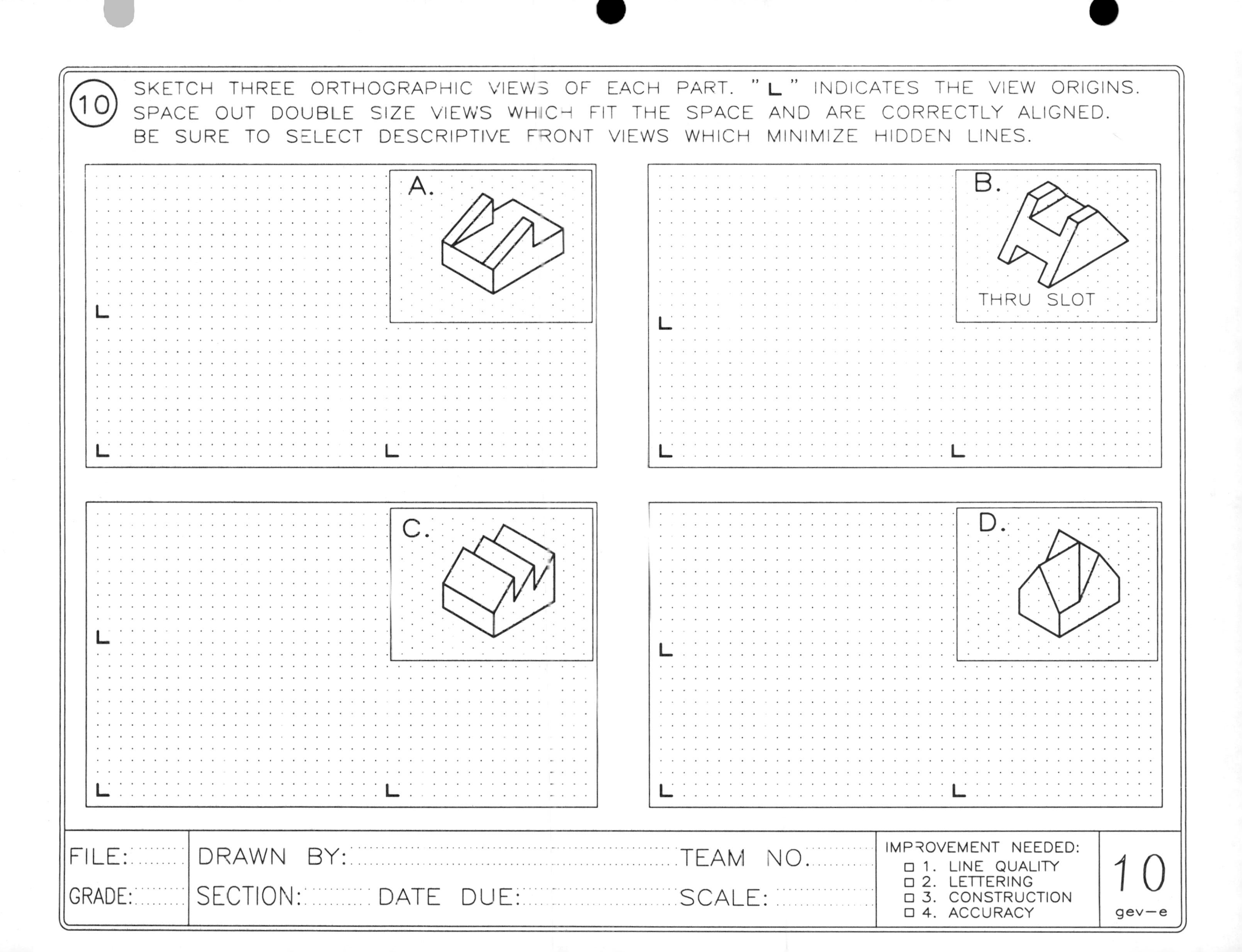
10
SKETCH THREE ORTHOGRAPHIC VIEWS OF EACH PART. "L" INDICATES THE VIEW ORIGINS.
SPACE OUT DOUBLE SIZE VIEWS WHICH FIT THE SPACE AND ARE CORRECTLY ALIGNED.
BE SURE TO SELECT DESCRIPTIVE FRONT VIEWS WHICH MINIMIZE HIDDEN LINES.
A.
B.
THRU SLOT
C.
D.
FILE:
GRADE:
DRAWN BY:
TEAM NO.
SECTION:
DATE DUE:
SCALE:
IMPROVEMENT NEEDED:
1. LINE QUALITY
2. LETTERING
3. CONSTRUCTION
4. ACCURACY
10
gev-e

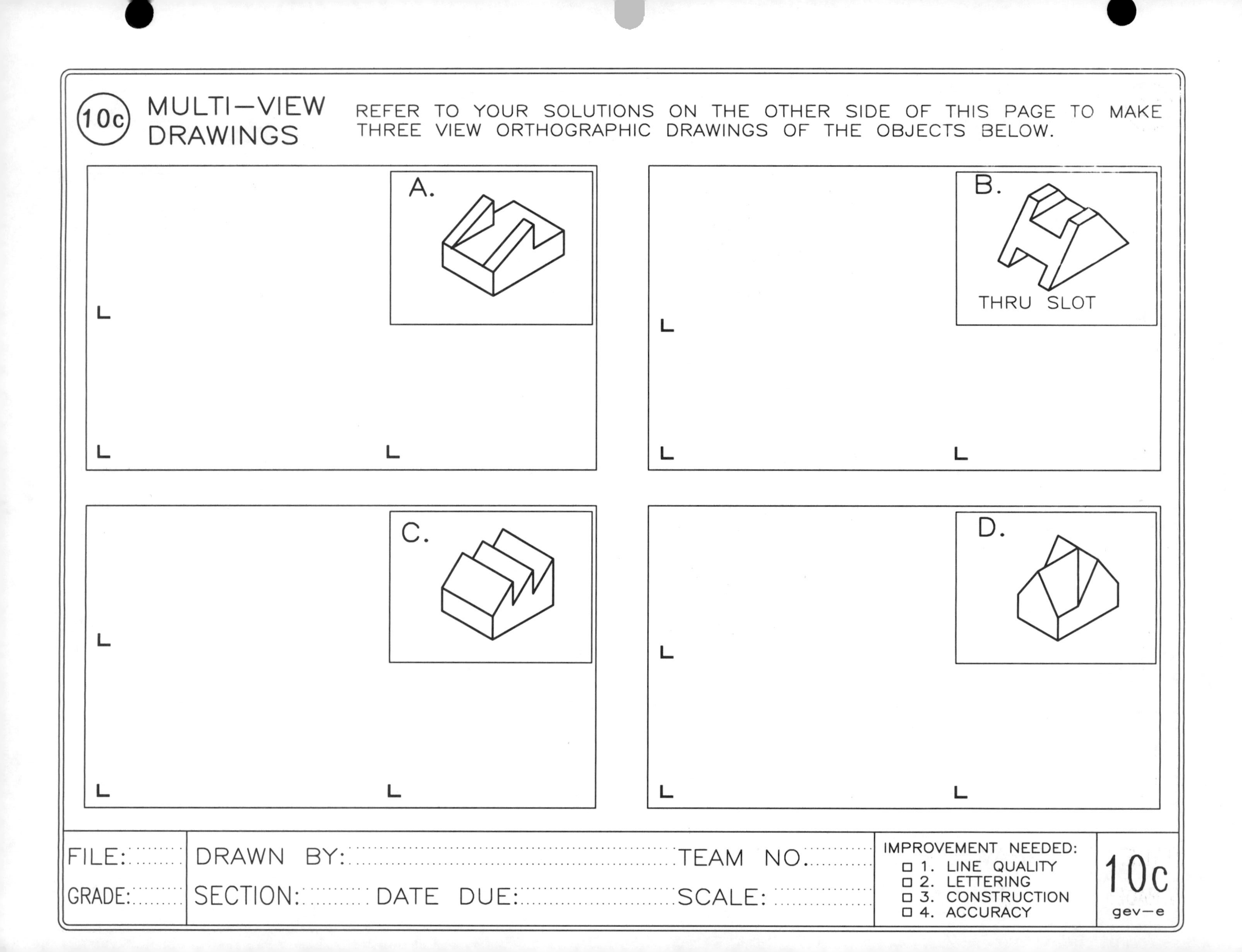
10c
MULTI—VIEW
DRAWINGS
REFER TO YOUR SOLUTIONS ON THE OTHER SIDE OF THIS PAGE TO MAKE
THREE VIEW ORTHOGRAPHIC DRAWINGS OF THE OBJECTS BELOW.
A.
B.
THRU SLOT
C.
D.
FILE:
GRADE:
DRAWN BY:
TEAM NO.
SECTION:
DATE DUE:
SCALE:
IMPROVEMENT NEEDED:
1. LINE QUALITY
2. LETTERING
3. CONSTRUCTION
4. ACCURACY
10c
gev—e

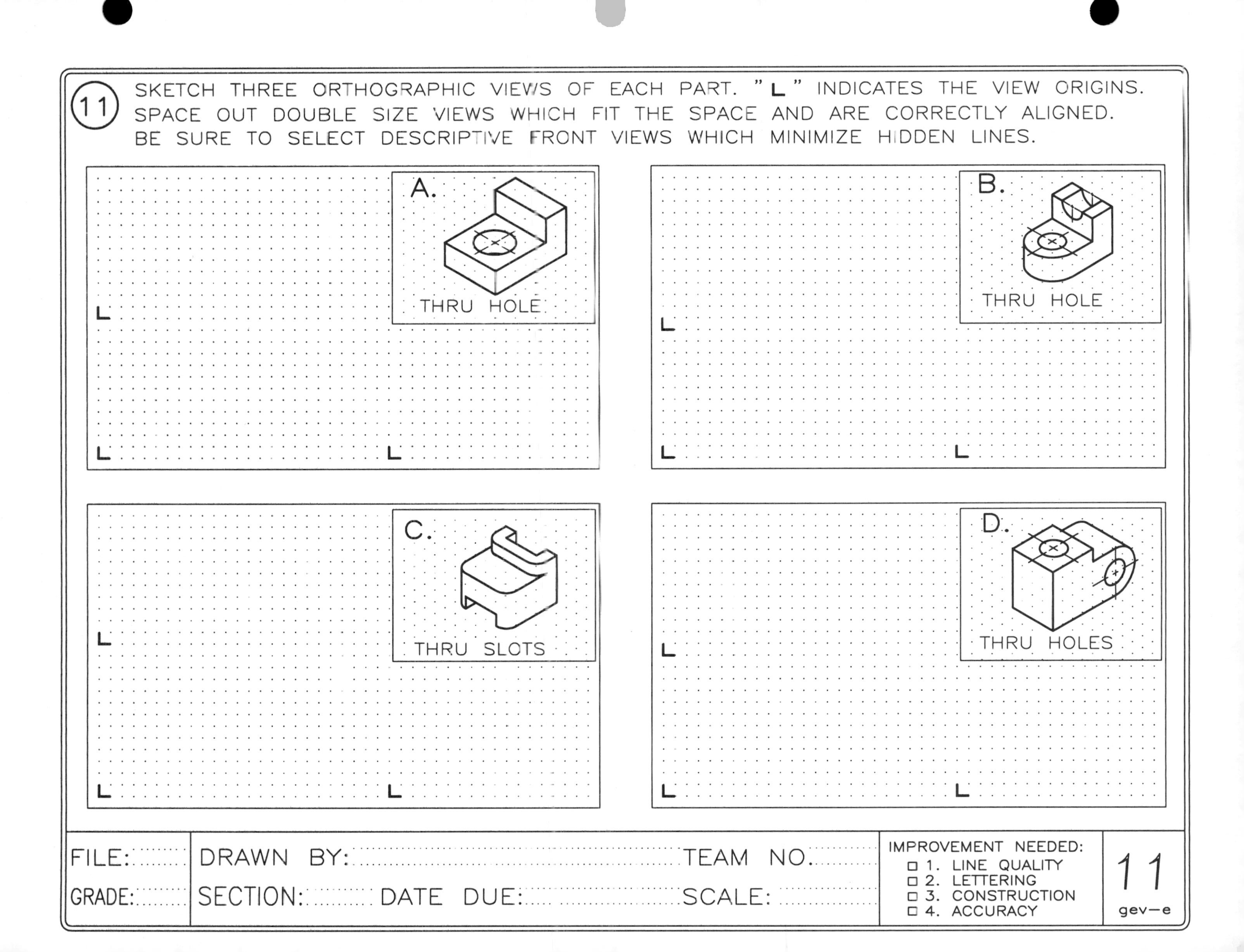
11
SKETCH THREE ORTHOGRAPHIC VIEWS OF EACH PART. "L" INDICATES THE VIEW ORIGINS.
SPACE OUT DOUBLE SIZE VIEWS WHICH FIT THE SPACE AND ARE CORRECTLY ALIGNED.
BE SURE TO SELECT DESCRIPTIVE FRONT VIEWS WHICH MINIMIZE HIDDEN LINES.
A.
THRU HOLE
B.
THRU HOLE
C.
THRU SLOTS
D.
THRU HOLES
FILE:
GRADE:
DRAWN BY:
TEAM NO.
SECTION:
DATE DUE:
SCALE:
IMPROVEMENT NEEDED:
1. LINE QUALITY
2. LETTERING
3. CONSTRUCTION
4. ACCURACY
11
gev-e

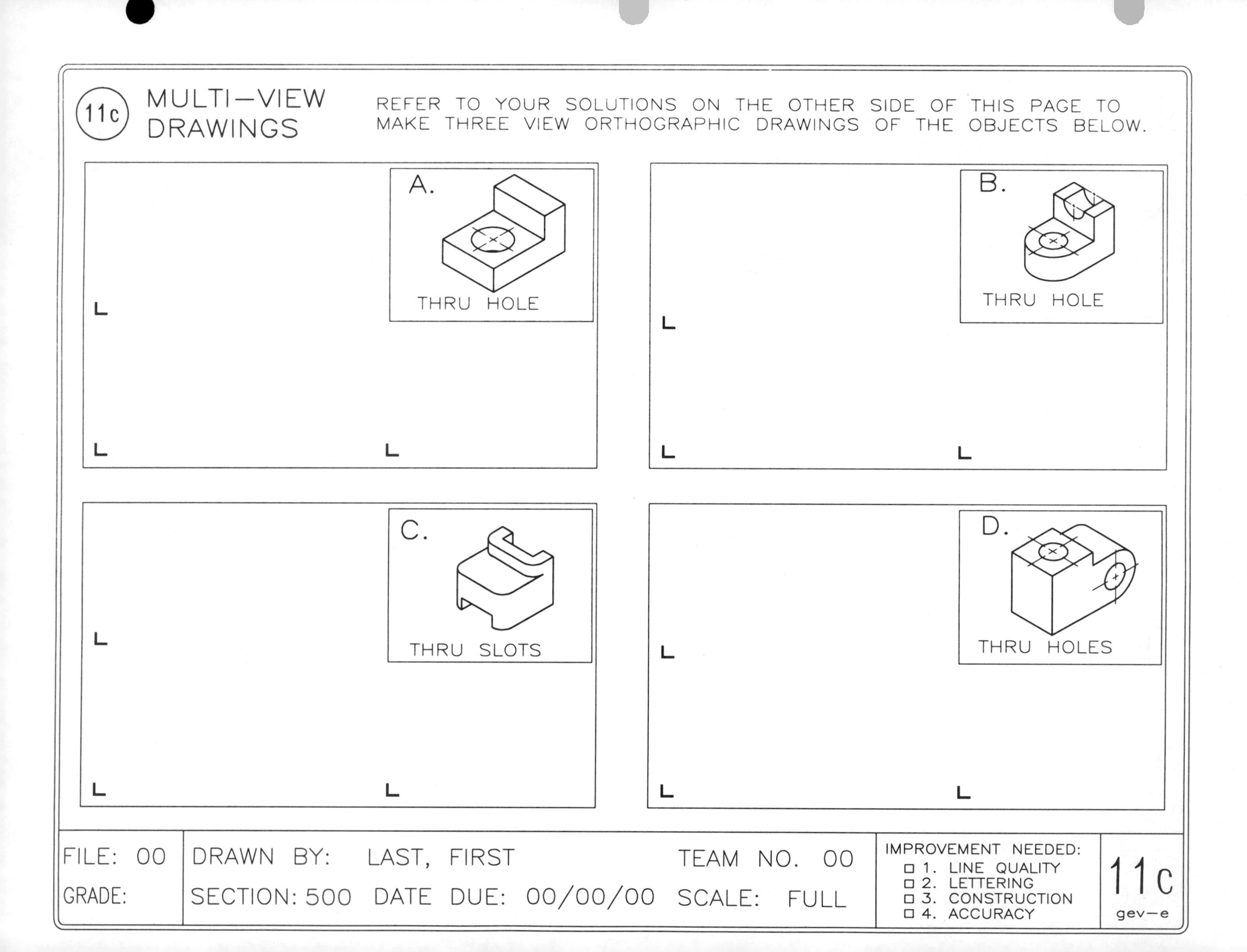
11c
MULTI—VIEW DRAWINGS
REFER TO YOUR SOLUTIONS ON THE OTHER SIDE OF THIS PAGE TO MAKE THREE VIEW ORTHOGRAPHIC DRAWINGS OF THE OBJECTS BELOW.
A.
THRU HOLE
B.
THRU HOLE
C.
THRU SLOTS
D.
THRU HOLES
FILE: 00
GRADE:
DRAWN BY: LAST, FIRST
TEAM NO. 00
SECTION: 500
DATE DUE: 00/00/00
SCALE: FULL
IMPROVEMENT NEEDED:
□ 1. LINE QUALITY
□ 2. LETTERING
□ 3. CONSTRUCTION
□ 4. ACCURACY
11c
gev—e

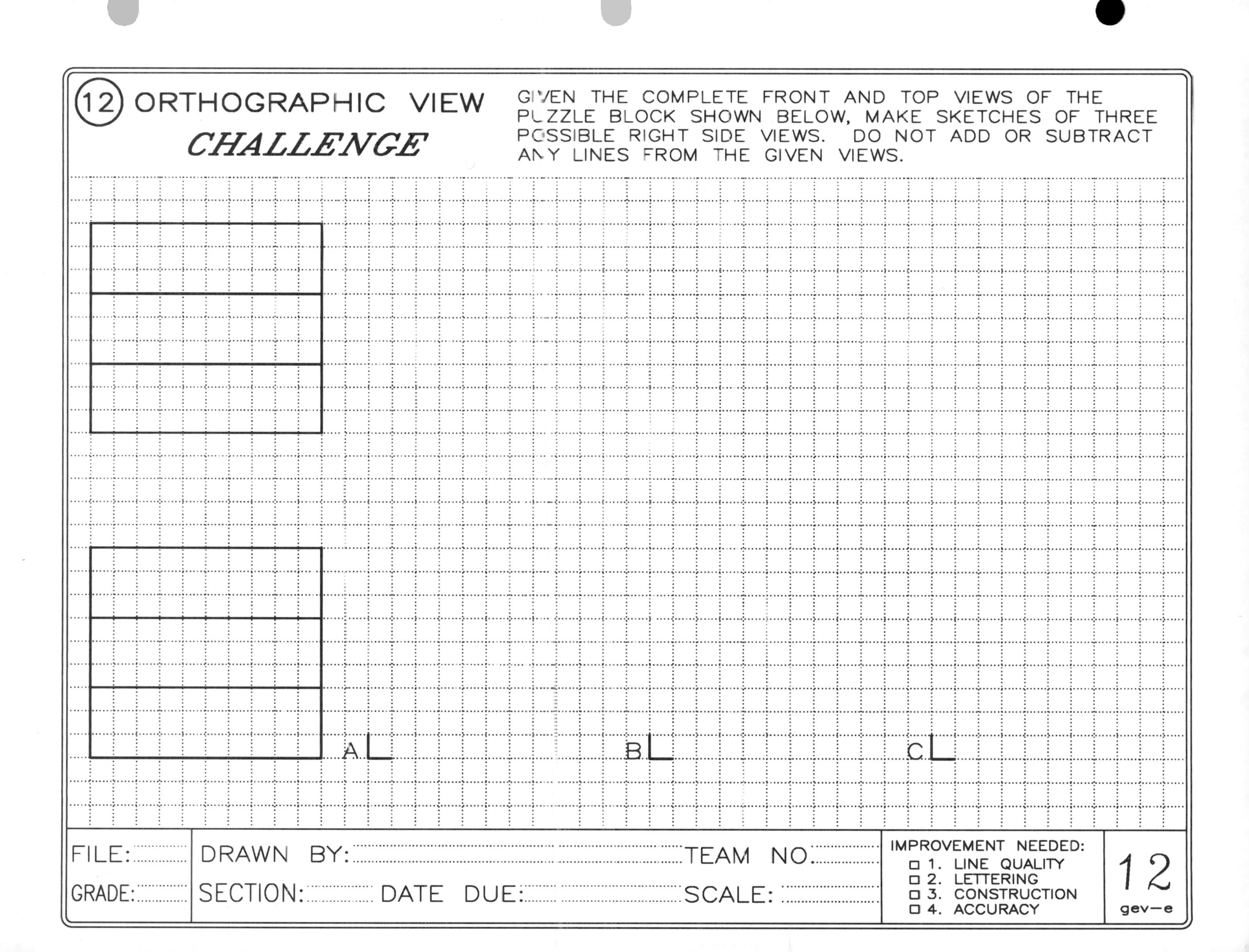
12
ORTHOGRAPHIC VIEW
CHALLENGE
GIVEN THE COMPLETE FRONT AND TOP VIEWS OF THE PUZZLE BLOCK SHOWN BELOW, MAKE SKETCHES OF THREE POSSIBLE RIGHT SIDE VIEWS. DO NOT ADD OR SUBTRACT ANY LINES FROM THE GIVEN VIEWS.
A
B
C
FILE:
GRADE:
DRAWN BY:
TEAM NO.
SECTION:
DATE DUE:
SCALE:
IMPROVEMENT NEEDED:
1. LINE QUALITY
2. LETTERING
3. CONSTRUCTION
4. ACCURACY
12
gev-e

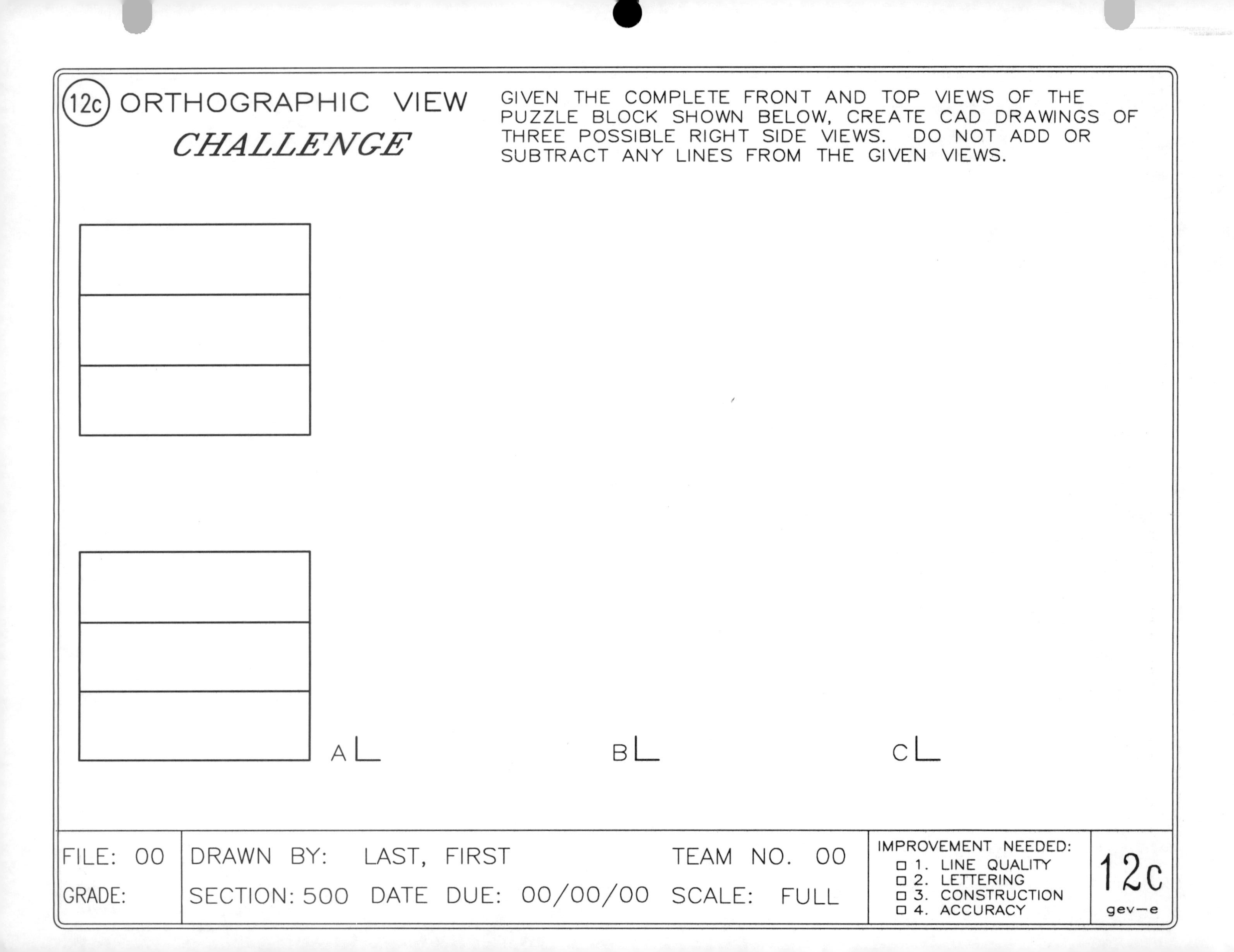
12c
ORTHOGRAPHIC VIEW
CHALLENGE
GIVEN THE COMPLETE FRONT AND TOP VIEWS OF THE PUZZLE BLOCK SHOWN BELOW, CREATE CAD DRAWINGS OF THREE POSSIBLE RIGHT SIDE VIEWS. DO NOT ADD OR SUBTRACT ANY LINES FROM THE GIVEN VIEWS.
A
B
C
FILE: 00
GRADE:
DRAWN BY: LAST, FIRST
TEAM NO. 00
SECTION: 500
DATE DUE: 00/00/00
SCALE: FULL
IMPROVEMENT NEEDED:
□ 1. LINE QUALITY
□ 2. LETTERING
□ 3. CONSTRUCTION
□ 4. ACCURACY
12c
gev—e

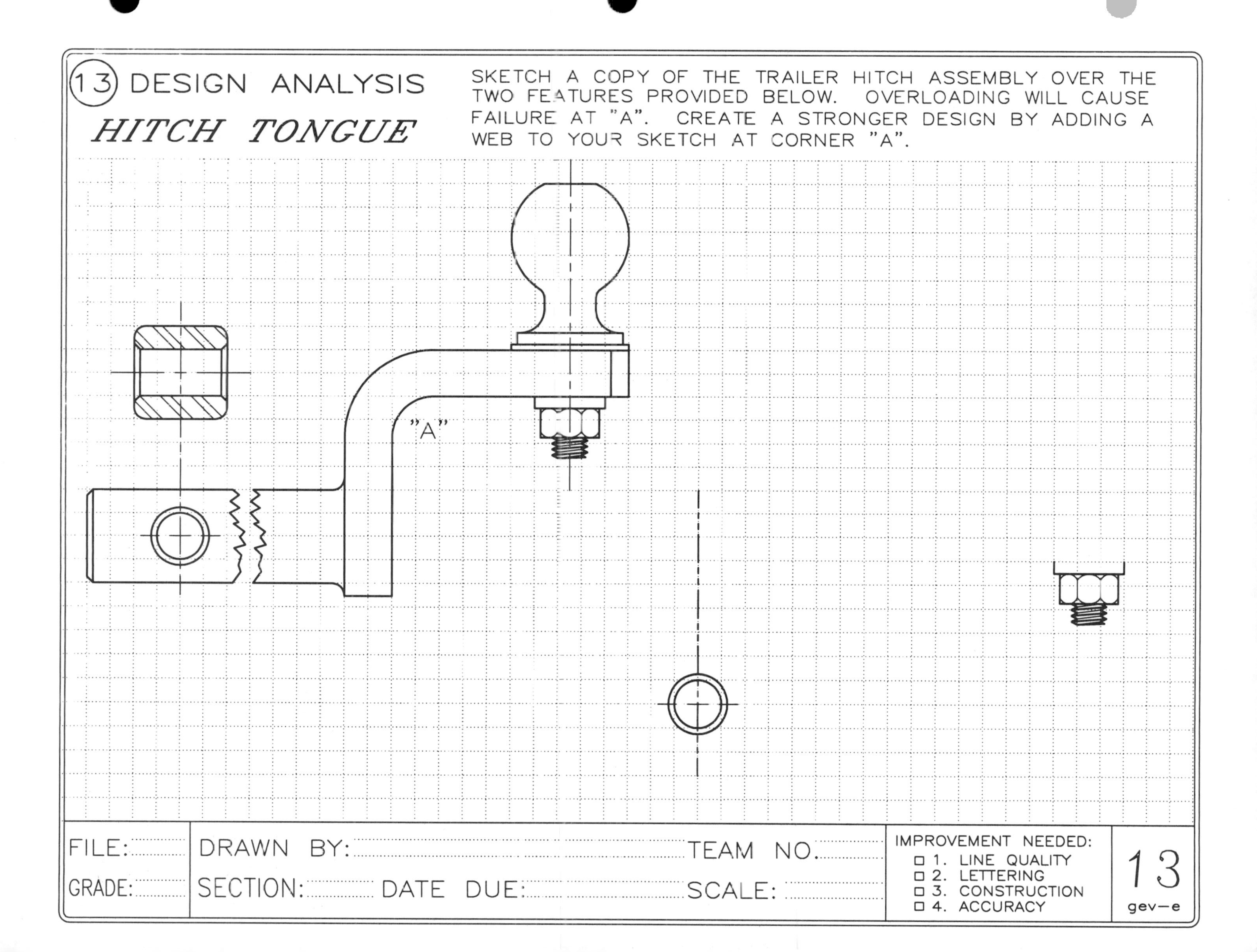
13 DESIGN ANALYSIS
HITCH TONGUE
SKETCH A COPY OF THE TRAILER HITCH ASSEMBLY OVER THE TWO FEATURES PROVIDED BELOW. OVERLOADING WILL CAUSE FAILURE AT "A". CREATE A STRONGER DESIGN BY ADDING A WEB TO YOUR SKETCH AT CORNER "A".
"A"
FILE:
GRADE:
DRAWN BY:
TEAM NO.
SECTION:
DATE DUE:
SCALE:
IMPROVEMENT NEEDED:
□ 1. LINE QUALITY
□ 2. LETTERING
□ 3. CONSTRUCTION
□ 4. ACCURACY
13
gev-e

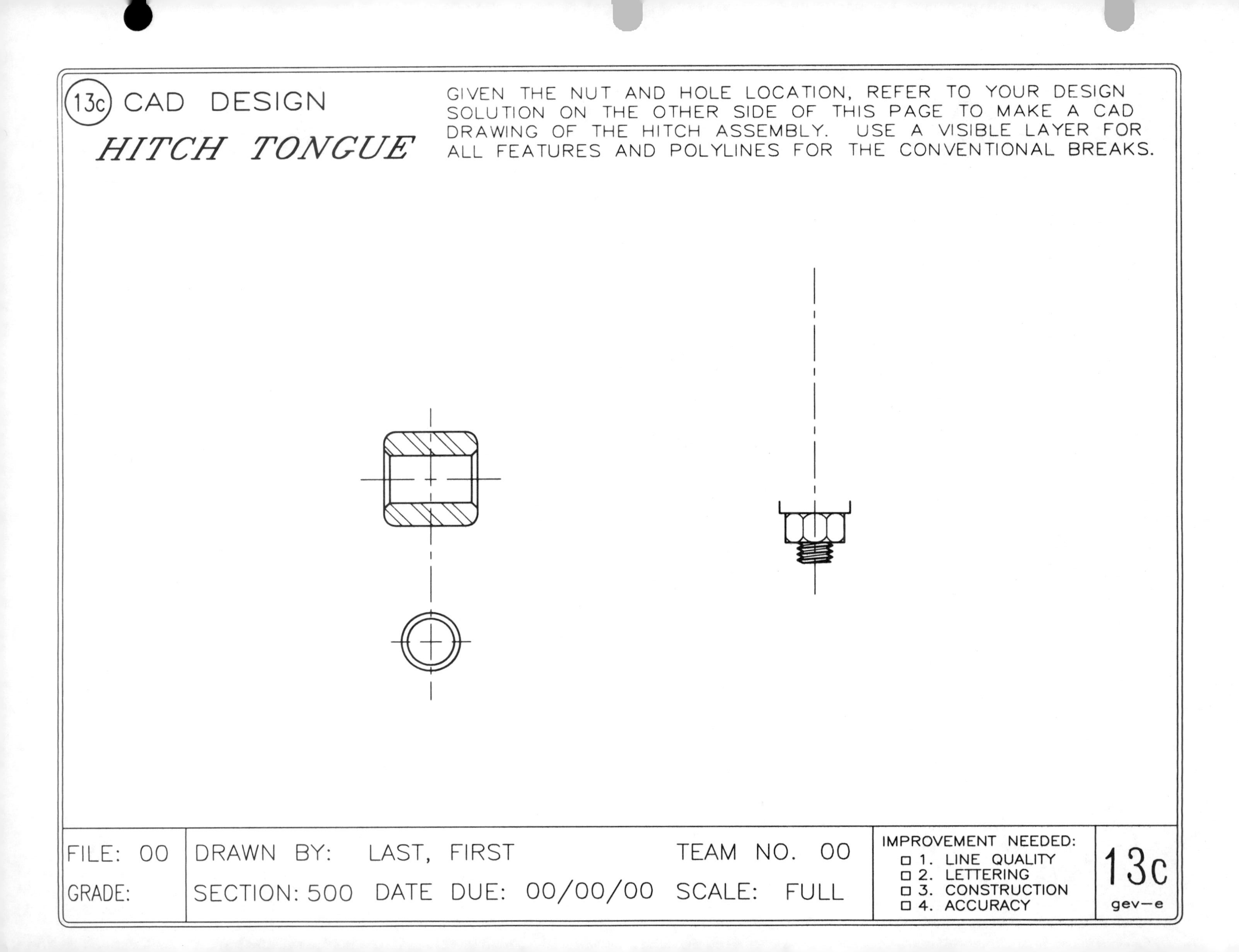
13c
CAD DESIGN
HITCH TONGUE
GIVEN THE NUT AND HOLE LOCATION, REFER TO YOUR DESIGN SOLUTION ON THE OTHER SIDE OF THIS PAGE TO MAKE A CAD DRAWING OF THE HITCH ASSEMBLY. USE A VISIBLE LAYER FOR ALL FEATURES AND POLYLINES FOR THE CONVENTIONAL BREAKS.
FILE: 00
GRADE:
DRAWN BY: LAST, FIRST
TEAM NO. 00
SECTION: 500
DATE DUE: 00/00/00
SCALE: FULL
IMPROVEMENT NEEDED:
□ 1. LINE QUALITY
□ 2. LETTERING
□ 3. CONSTRUCTION
□ 4. ACCURACY
13c
gev—e

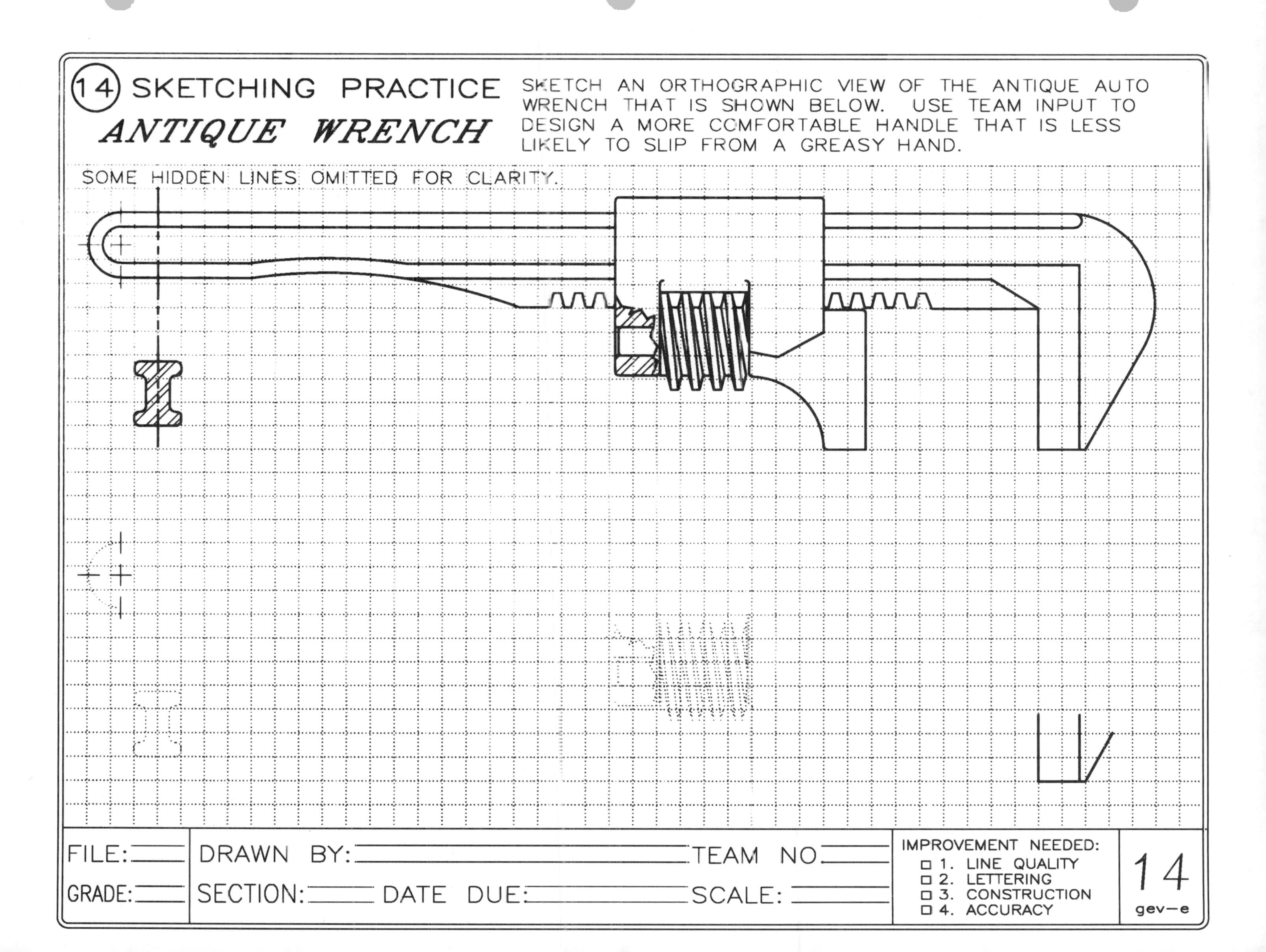

(14) SKETCHING PRACTICE
ANTIQUE WRENCH
SKETCH AN ORTHOGRAPHIC VIEW OF THE ANTIQUE AUTO WRENCH THAT IS SHOWN BELOW. USE TEAM INPUT TO DESIGN A MORE COMFORTABLE HANDLE THAT IS LESS LIKELY TO SLIP FROM A GREASY HAND.
SOME HIDDEN LINES OMITTED FOR CLARITY.
FILE:
GRADE:
DRAWN BY:
TEAM NO
SECTION:
DATE DUE:
SCALE:
IMPROVEMENT NEEDED:
□ 1. LINE QUALITY
□ 2. LETTERING
□ 3. CONSTRUCTION
□ 4. ACCURACY
14
gev-e

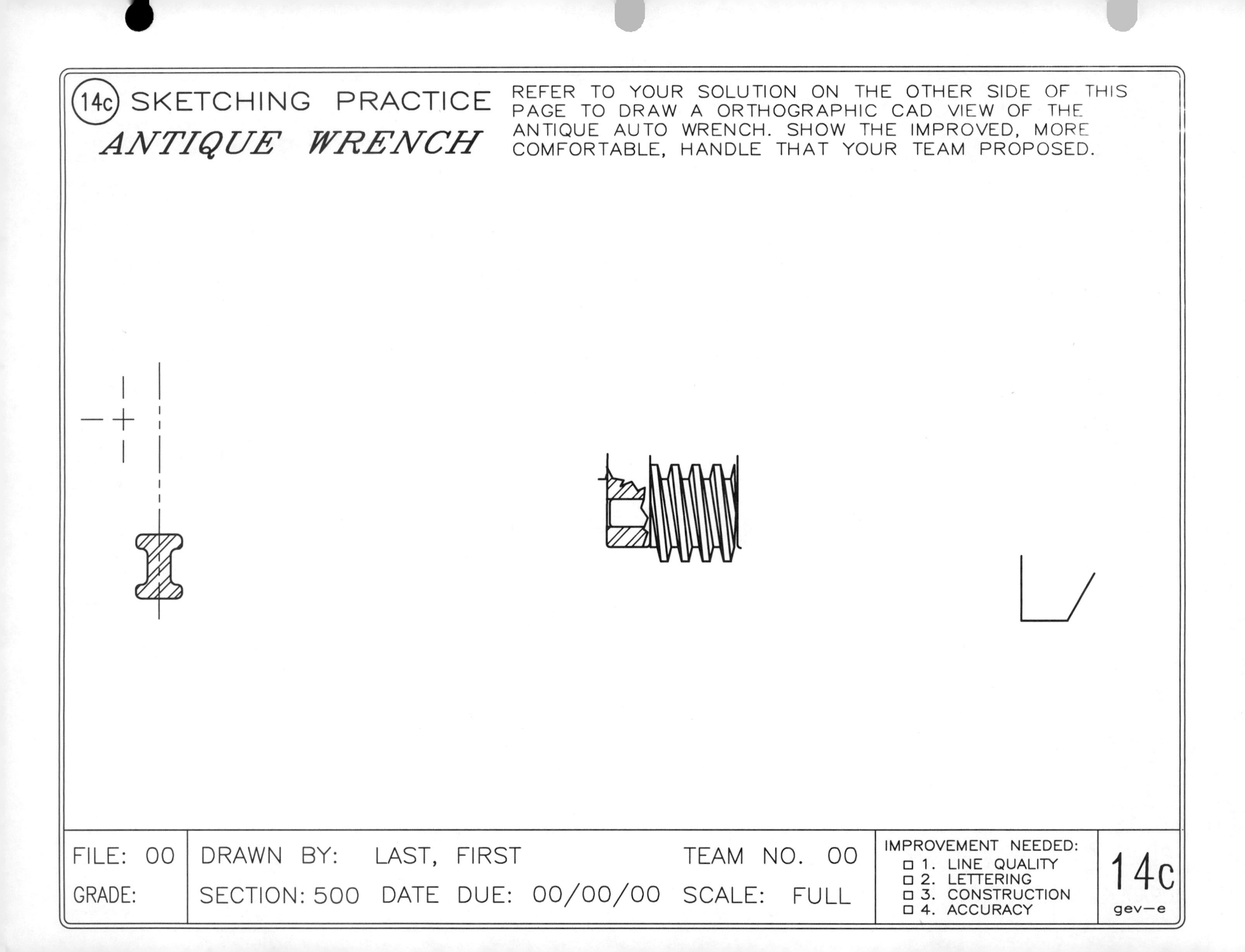

14c
SKETCHING PRACTICE
ANTIQUE WRENCH
REFER TO YOUR SOLUTION ON THE OTHER SIDE OF THIS PAGE TO DRAW A ORTHOGRAPHIC CAD VIEW OF THE ANTIQUE AUTO WRENCH. SHOW THE IMPROVED, MORE COMFORTABLE, HANDLE THAT YOUR TEAM PROPOSED.
FILE: 00
GRADE:
DRAWN BY: LAST, FIRST
TEAM NO. 00
SECTION: 500
DATE DUE: 00/00/00
SCALE: FULL
IMPROVEMENT NEEDED:
□ 1. LINE QUALITY
□ 2. LETTERING
□ 3. CONSTRUCTION
□ 4. ACCURACY
14c
gev—e

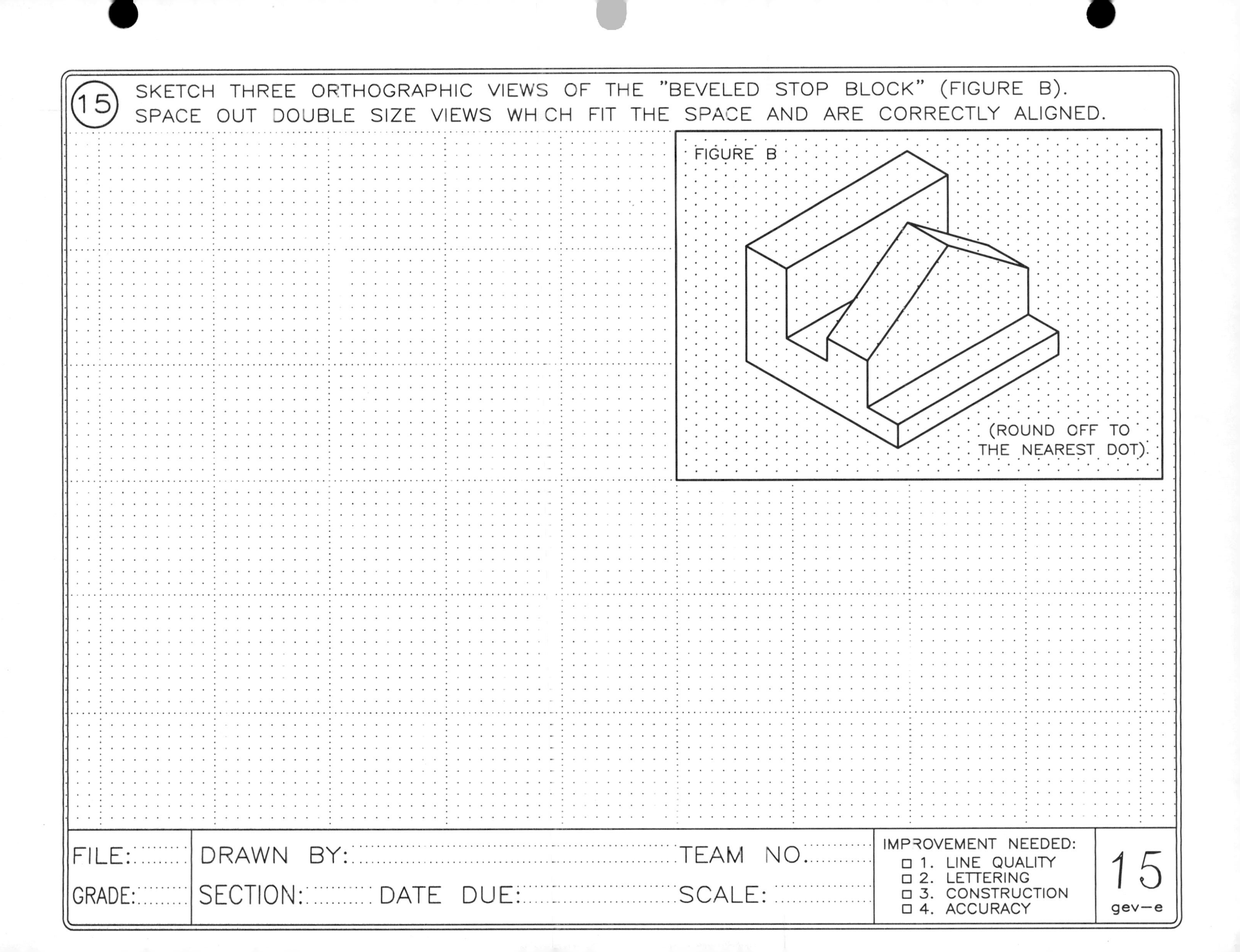

(15) SKETCH THREE ORTHOGRAPHIC VIEWS OF THE "BEVELED STOP BLOCK" (FIGURE B).
SPACE OUT DOUBLE SIZE VIEWS WHICH FIT THE SPACE AND ARE CORRECTLY ALIGNED.

FILE:	DRAWN BY:	TEAM NO.	IMPROVEMENT NEEDED: ☐ 1. LINE QUALITY ☐ 2. LETTERING ☐ 3. CONSTRUCTION ☐ 4. ACCURACY	15
GRADE:	SECTION: DATE DUE:	SCALE:		gev–e

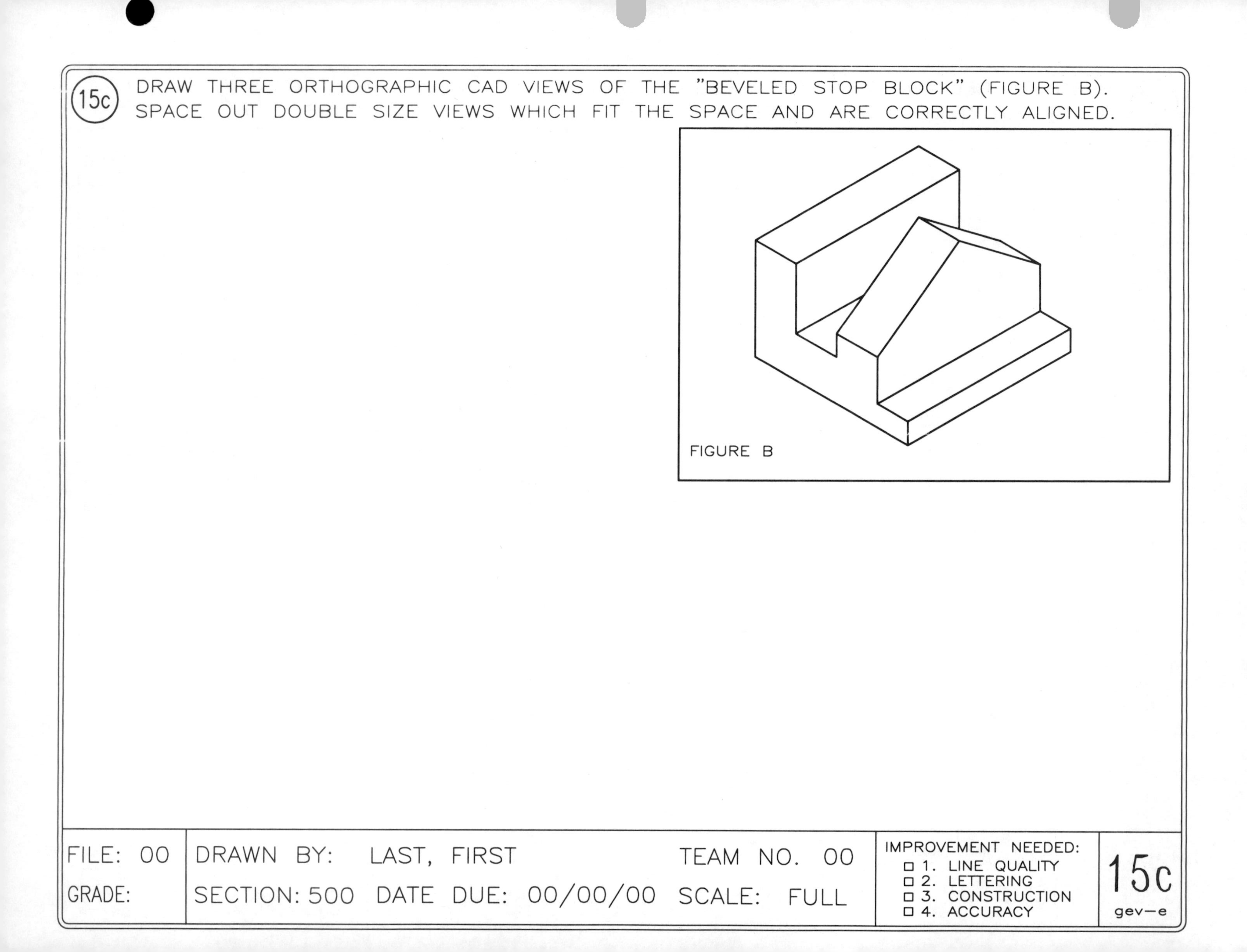
15c
DRAW THREE ORTHOGRAPHIC CAD VIEWS OF THE "BEVELED STOP BLOCK" (FIGURE B).
SPACE OUT DOUBLE SIZE VIEWS WHICH FIT THE SPACE AND ARE CORRECTLY ALIGNED.
FIGURE B
FILE: 00
GRADE:
DRAWN BY: LAST, FIRST
TEAM NO. 00
SECTION: 500
DATE DUE: 00/00/00
SCALE: FULL
IMPROVEMENT NEEDED:
□ 1. LINE QUALITY
□ 2. LETTERING
□ 3. CONSTRUCTION
□ 4. ACCURACY
15c
gev-e

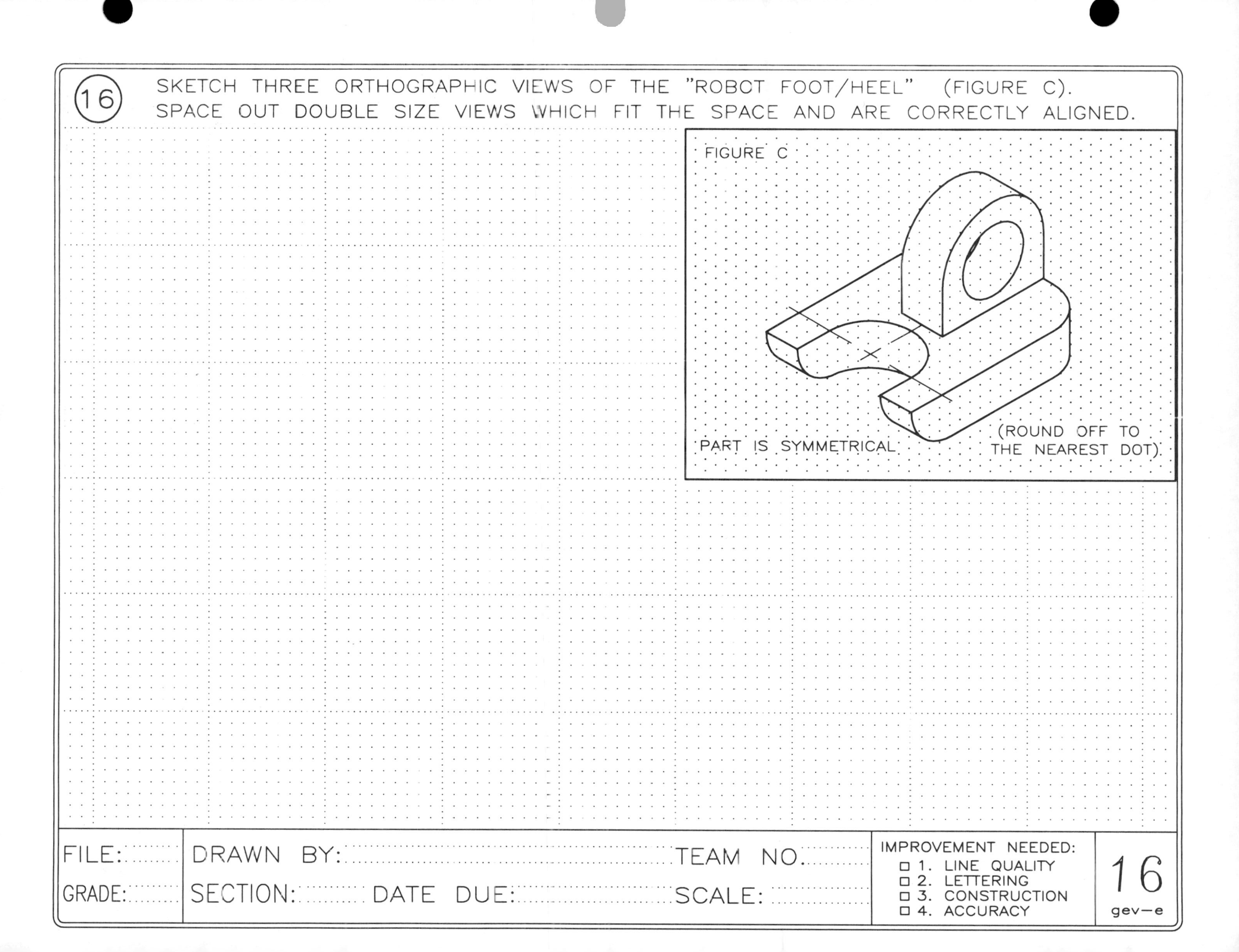

16
SKETCH THREE ORTHOGRAPHIC VIEWS OF THE "ROBOT FOOT/HEEL" (FIGURE C).
SPACE OUT DOUBLE SIZE VIEWS WHICH FIT THE SPACE AND ARE CORRECTLY ALIGNED.
FIGURE C
PART IS SYMMETRICAL
(ROUND OFF TO THE NEAREST DOT)
FILE:
GRADE:
DRAWN BY:
TEAM NO.
SECTION:
DATE DUE:
SCALE:
IMPROVEMENT NEEDED:
☐ 1. LINE QUALITY
☐ 2. LETTERING
☐ 3. CONSTRUCTION
☐ 4. ACCURACY
16
gev—e

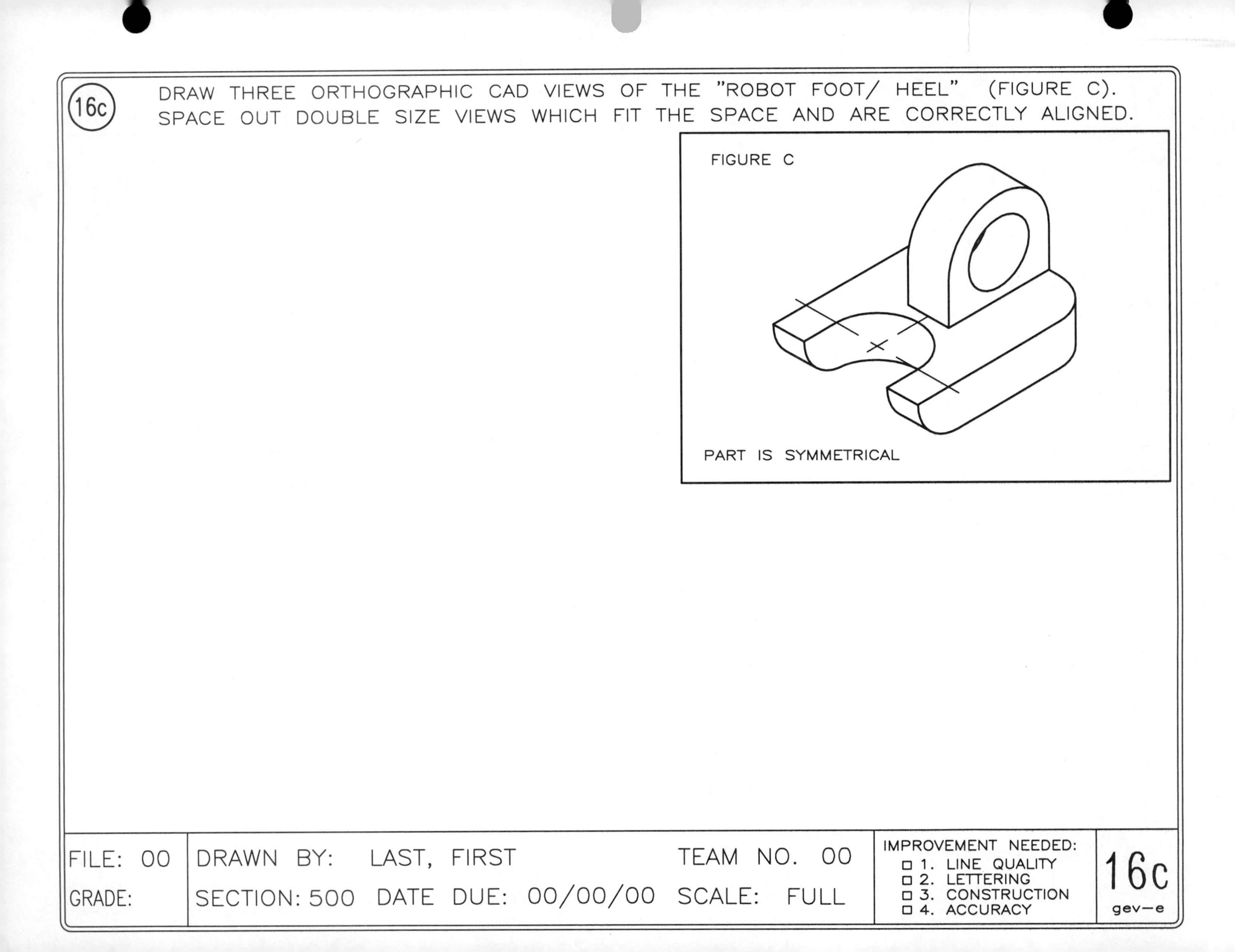
16c
DRAW THREE ORTHOGRAPHIC CAD VIEWS OF THE "ROBOT FOOT/ HEEL" (FIGURE C).
SPACE OUT DOUBLE SIZE VIEWS WHICH FIT THE SPACE AND ARE CORRECTLY ALIGNED.
FIGURE C
PART IS SYMMETRICAL
FILE: 00
GRADE:
DRAWN BY: LAST, FIRST
TEAM NO. 00
SECTION: 500 DATE DUE: 00/00/00 SCALE: FULL
IMPROVEMENT NEEDED:
□ 1. LINE QUALITY
□ 2. LETTERING
□ 3. CONSTRUCTION
□ 4. ACCURACY
16c
gev—e

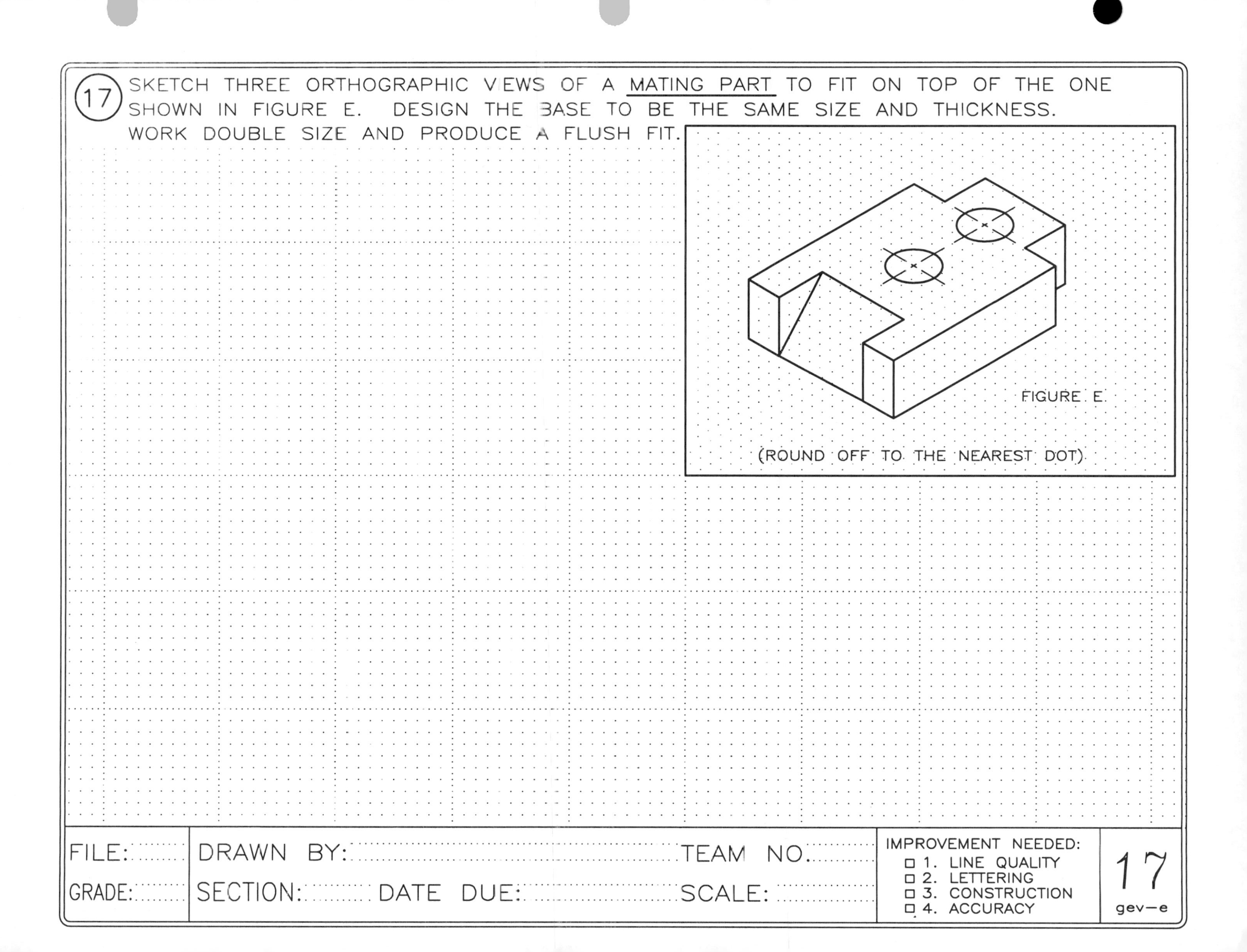
17
SKETCH THREE ORTHOGRAPHIC VIEWS OF A MATING PART TO FIT ON TOP OF THE ONE SHOWN IN FIGURE E. DESIGN THE BASE TO BE THE SAME SIZE AND THICKNESS. WORK DOUBLE SIZE AND PRODUCE A FLUSH FIT.
FIGURE E.
(ROUND OFF TO THE NEAREST DOT)
FILE:
GRADE:
DRAWN BY:
SECTION:
DATE DUE:
TEAM NO.
SCALE:
IMPROVEMENT NEEDED:
1. LINE QUALITY
2. LETTERING
3. CONSTRUCTION
4. ACCURACY
17
gev—e

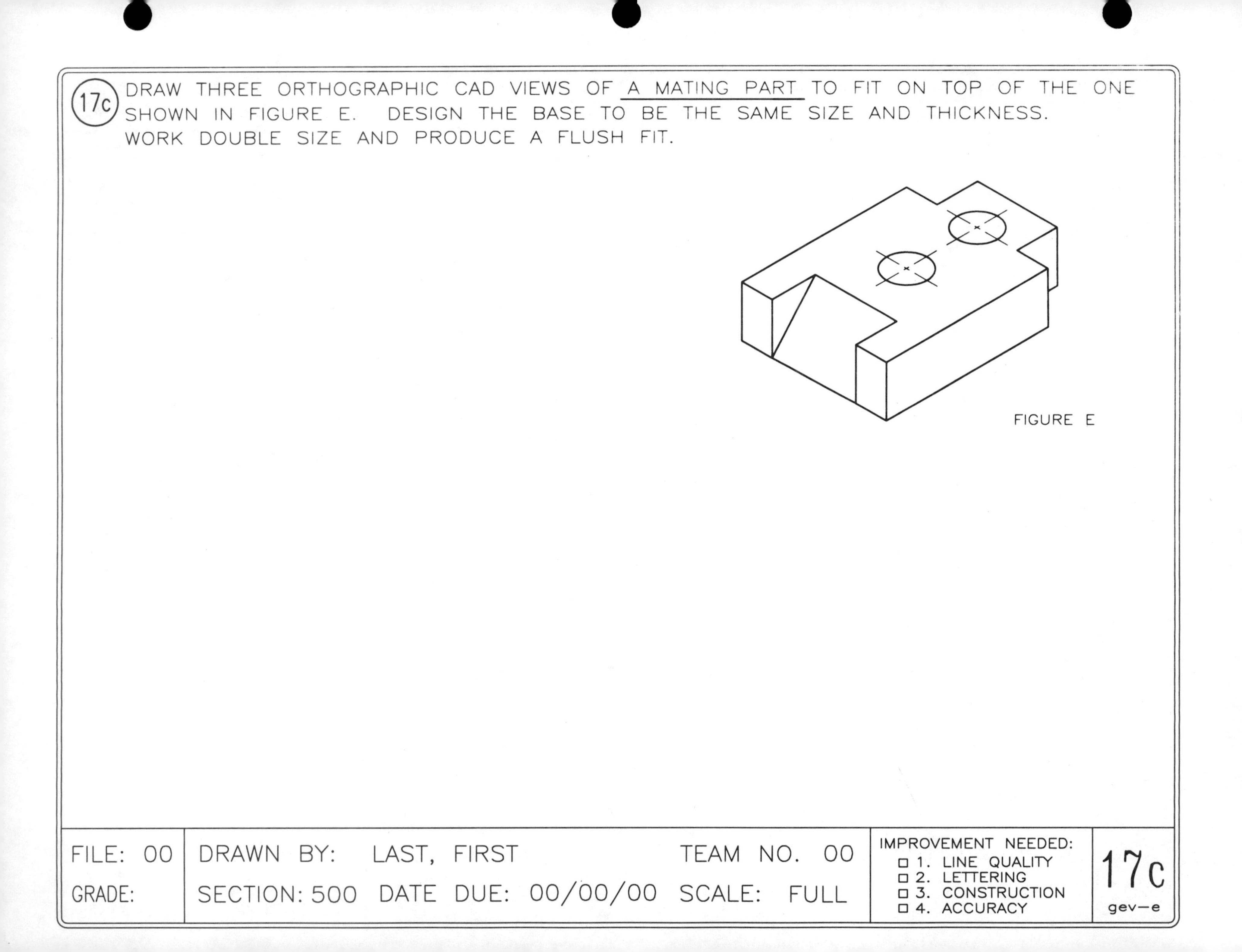

(17c) DRAW THREE ORTHOGRAPHIC CAD VIEWS OF A MATING PART TO FIT ON TOP OF THE ONE SHOWN IN FIGURE E. DESIGN THE BASE TO BE THE SAME SIZE AND THICKNESS. WORK DOUBLE SIZE AND PRODUCE A FLUSH FIT.

FIGURE E

FILE: 00	DRAWN BY: LAST, FIRST	TEAM NO. 00	IMPROVEMENT NEEDED:	17c
GRADE:	SECTION: 500 DATE DUE: 00/00/00	SCALE: FULL	☐ 1. LINE QUALITY ☐ 2. LETTERING ☐ 3. CONSTRUCTION ☐ 4. ACCURACY	gev–e

(18) SKETCH THREE CRTHOGRAPHIC VIEWS OF THE "S" LINK SHOWN IN FIGURE F.
USE A FULL SCALE TO FIT THE SPACE AND CORRECTLY ALIGN THE VIEWS.

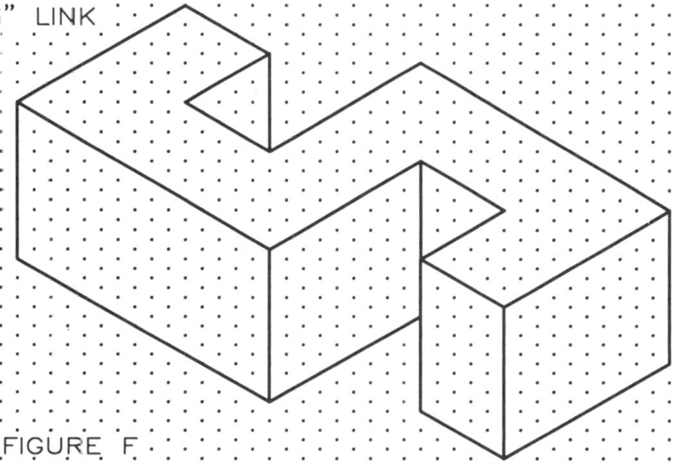

FIGURE F

(ROUND OFF TO THE NEAREST DOT)

ANALYZE THE LINK TO SEE IF IT WILL MAKE A FLUSH FIT WITH AN IDENTICAL ONE.

FILE:
GRADE:
DRAWN BY: TEAM NO.
SECTION: DATE DUE: SCALE:

IMPROVEMENT NEEDED:
- ☐ 1. LINE QUALITY
- ☐ 2. LETTERING
- ☐ 3. CONSTRUCTION
- ☐ 4. ACCURACY

18
gev—e

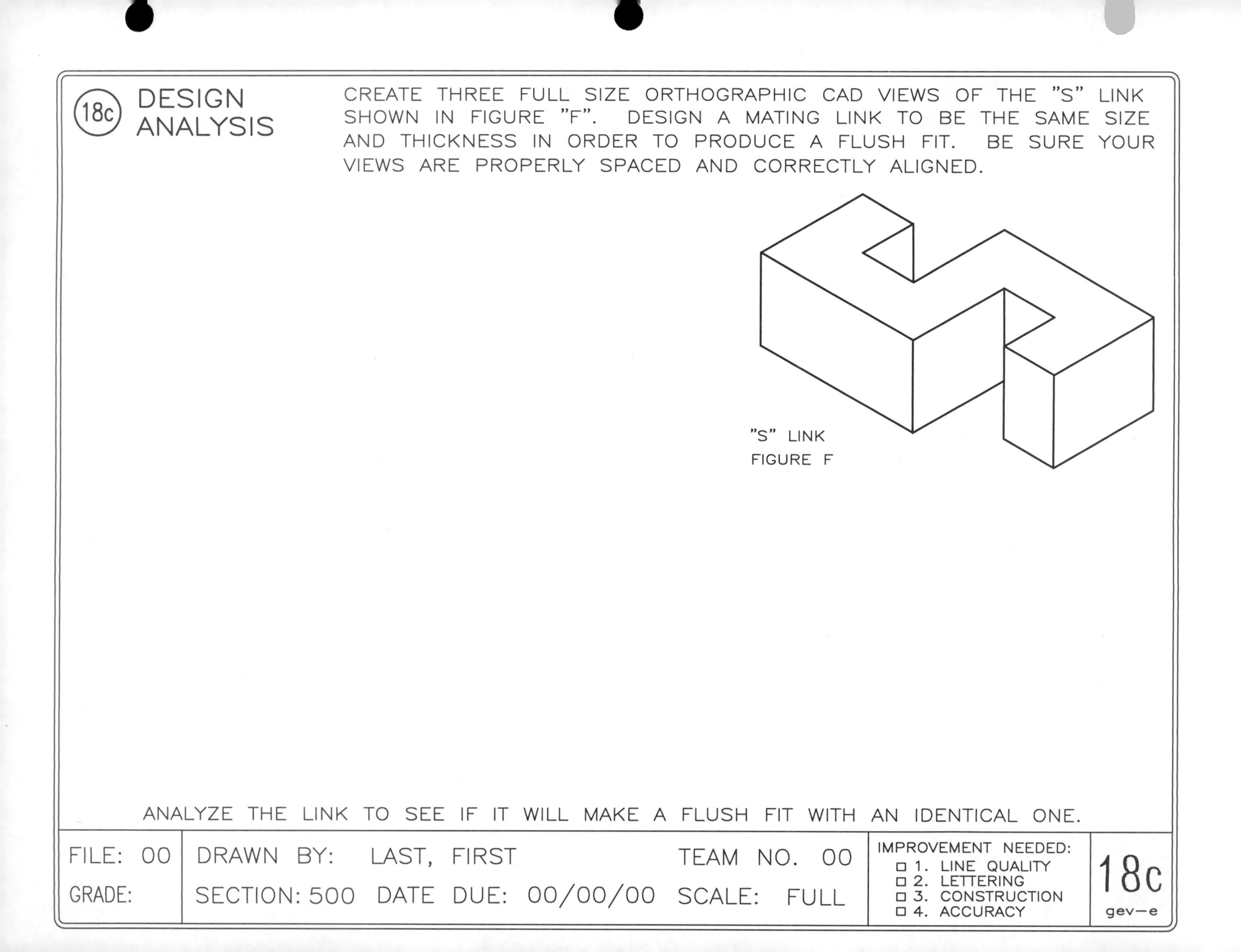

18c DESIGN ANALYSIS

CREATE THREE FULL SIZE ORTHOGRAPHIC CAD VIEWS OF THE "S" LINK SHOWN IN FIGURE "F". DESIGN A MATING LINK TO BE THE SAME SIZE AND THICKNESS IN ORDER TO PRODUCE A FLUSH FIT. BE SURE YOUR VIEWS ARE PROPERLY SPACED AND CORRECTLY ALIGNED.

"S" LINK
FIGURE F

ANALYZE THE LINK TO SEE IF IT WILL MAKE A FLUSH FIT WITH AN IDENTICAL ONE.

FILE: 00	DRAWN BY: LAST, FIRST	TEAM NO. 00	IMPROVEMENT NEEDED: ☐ 1. LINE QUALITY ☐ 2. LETTERING ☐ 3. CONSTRUCTION ☐ 4. ACCURACY	18c
GRADE:	SECTION: 500 DATE DUE: 00/00/00	SCALE: FULL		gev—e

(19) COMPLETE THE ORTHOGRAPHIC VIEWS OF THE PARTS SHOWN BELOW. LINES MAY BE MISSING FROM ANY VIEW. SKETCH HALF SIZE PICTORIALS OF EACH PART IN THE BOX PROVIDED. THIS WILL HELP YOU VISUALIZE ANY MISSING LINES.

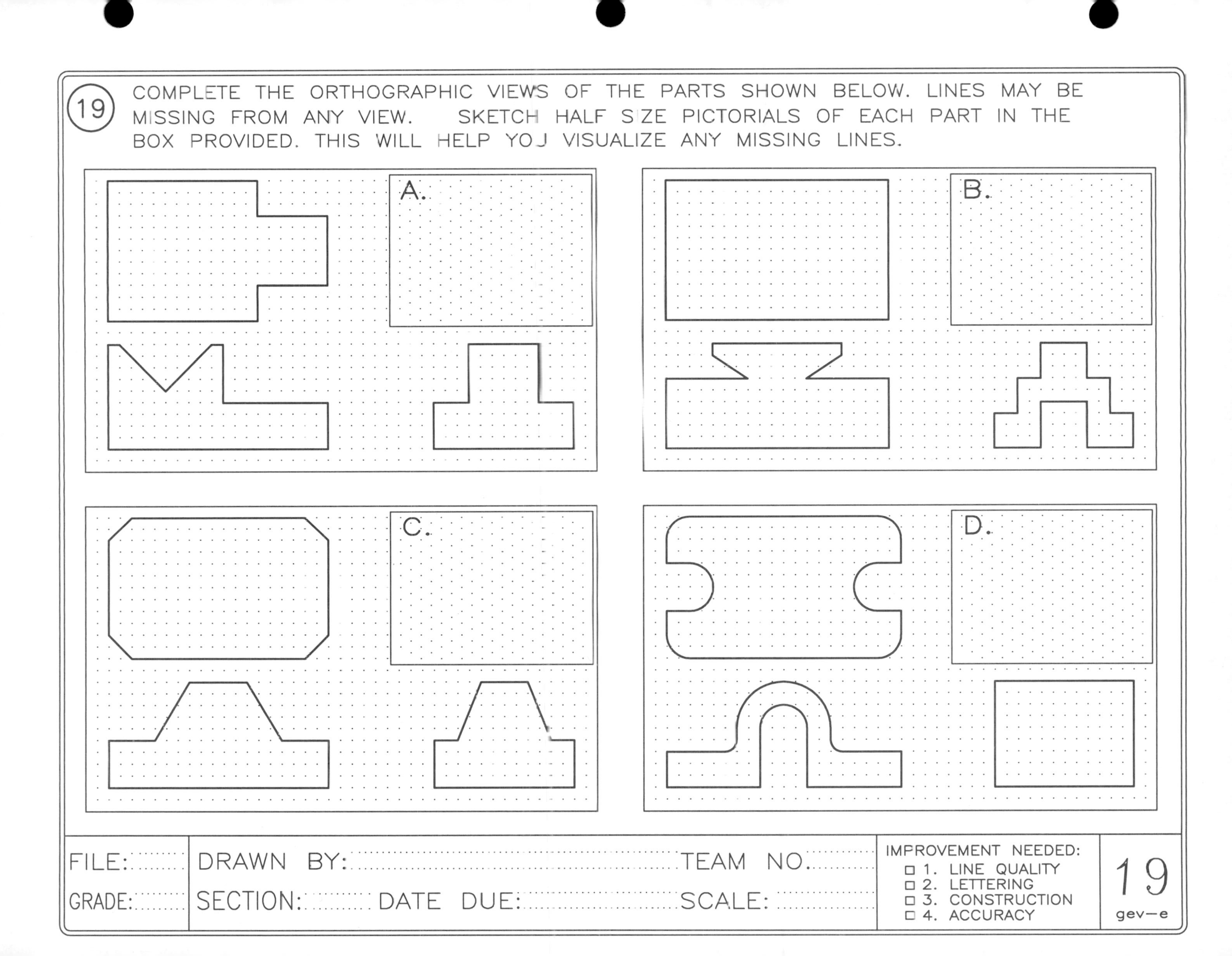

FILE:	DRAWN BY:	TEAM NO.	IMPROVEMENT NEEDED:	19
GRADE:	SECTION: DATE DUE:	SCALE:	☐ 1. LINE QUALITY ☐ 2. LETTERING ☐ 3. CONSTRUCTION ☐ 4. ACCURACY	gev—e

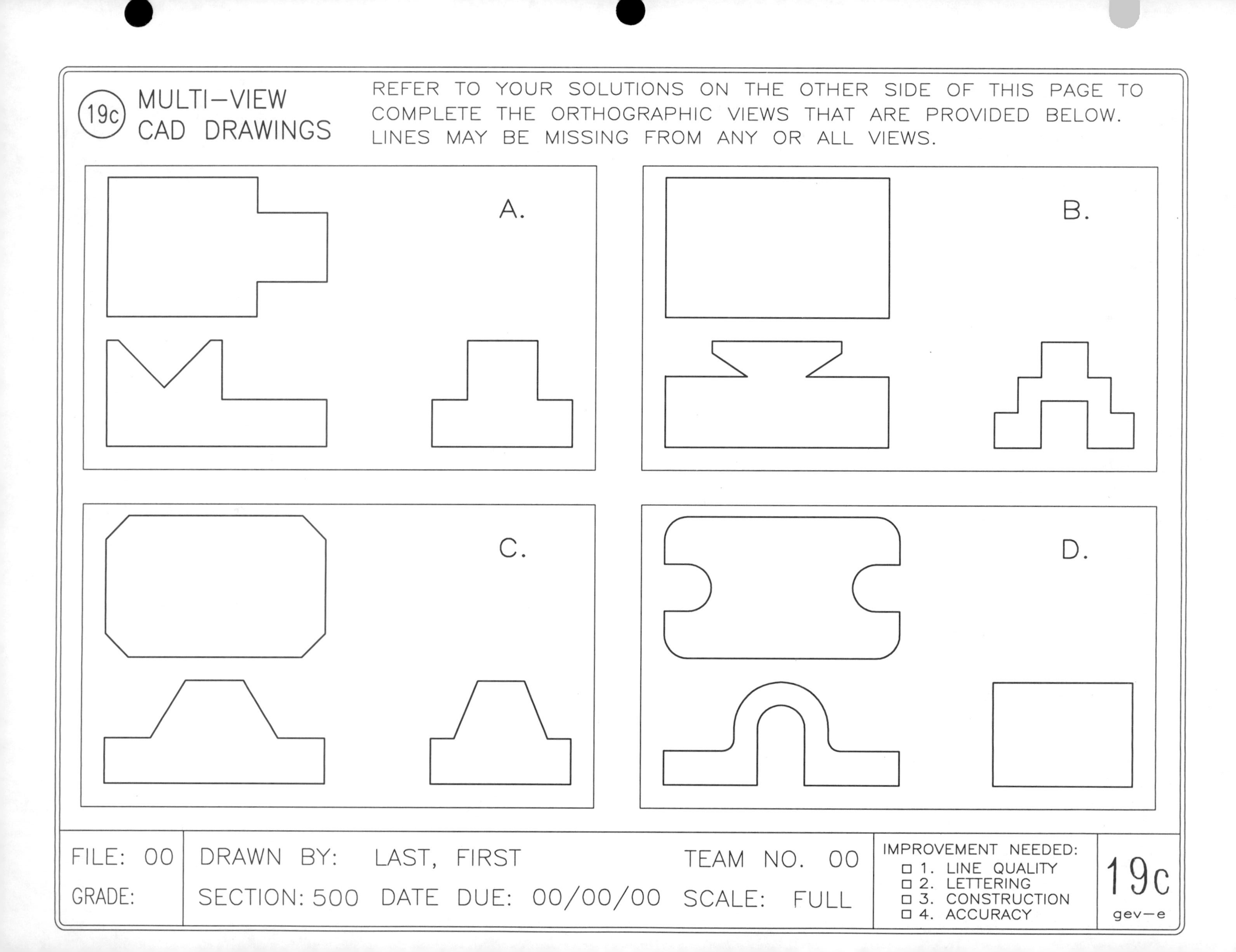

19c
MULTI–VIEW
CAD DRAWINGS
REFER TO YOUR SOLUTIONS ON THE OTHER SIDE OF THIS PAGE TO COMPLETE THE ORTHOGRAPHIC VIEWS THAT ARE PROVIDED BELOW. LINES MAY BE MISSING FROM ANY OR ALL VIEWS.
A.
B.
C.
D.
FILE: 00
GRADE:
DRAWN BY: LAST, FIRST
TEAM NO. 00
SECTION: 500
DATE DUE: 00/00/00
SCALE: FULL
IMPROVEMENT NEEDED:
□ 1. LINE QUALITY
□ 2. LETTERING
□ 3. CONSTRUCTION
□ 4. ACCURACY
19c
gev–e

(20) COMPLETE THE ORTHOGRAPHIC VIEWS OF THE PARTS SHOWN BELOW. LINES MAY BE MISSING FROM ANY VIEW. SKETCH HALF SIZE PICTORIALS OF EACH PART IN THE BOX PROVIDED. THIS WILL HELP YOU VISUALIZE ANY MISSING LINES.

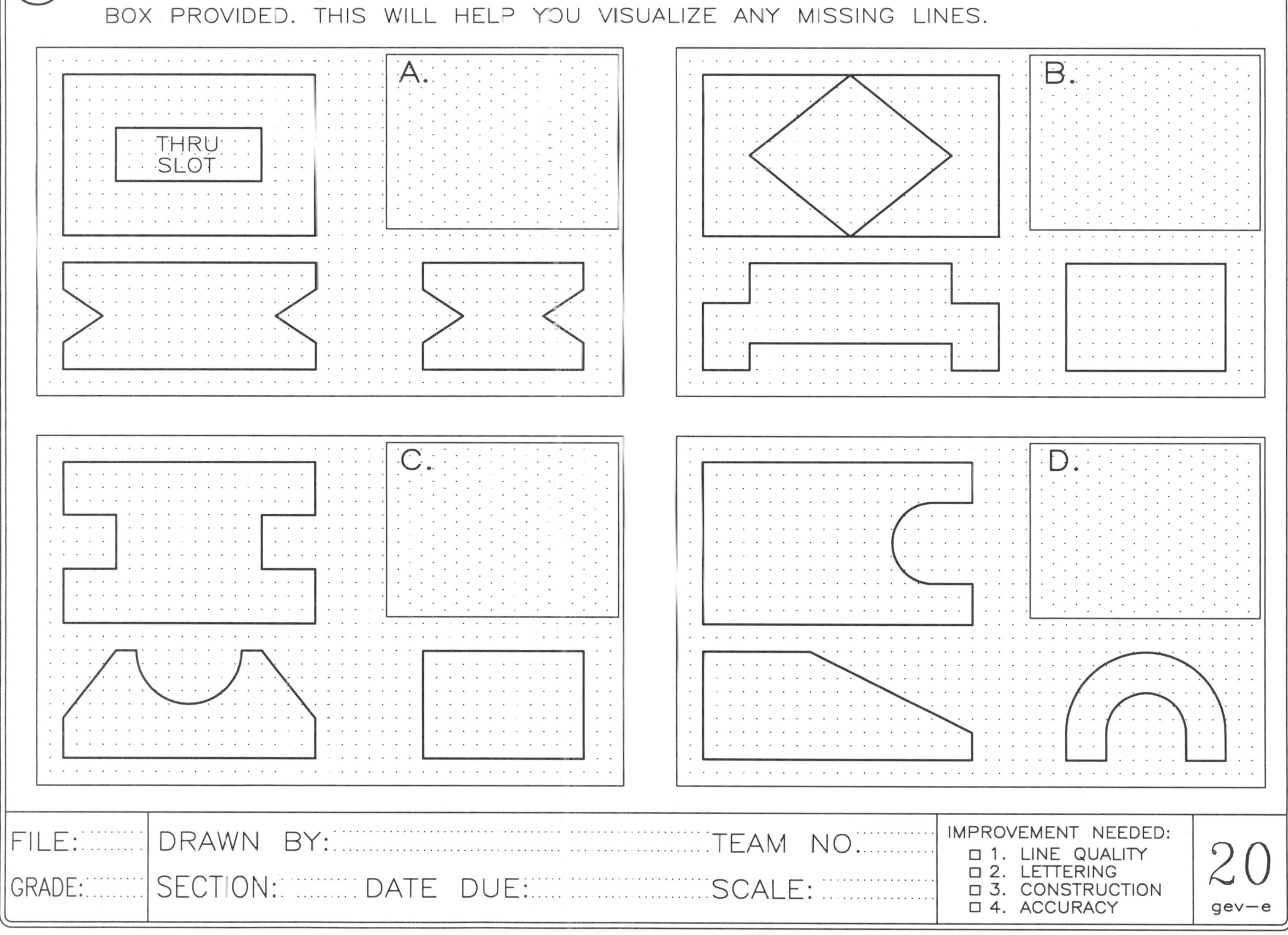

FILE: ..... DRAWN BY: ..... TEAM NO. .....

GRADE: ..... SECTION: ..... DATE DUE: ..... SCALE: .....

IMPROVEMENT NEEDED:
- ☐ 1. LINE QUALITY
- ☐ 2. LETTERING
- ☐ 3. CONSTRUCTION
- ☐ 4. ACCURACY

20

gev–e

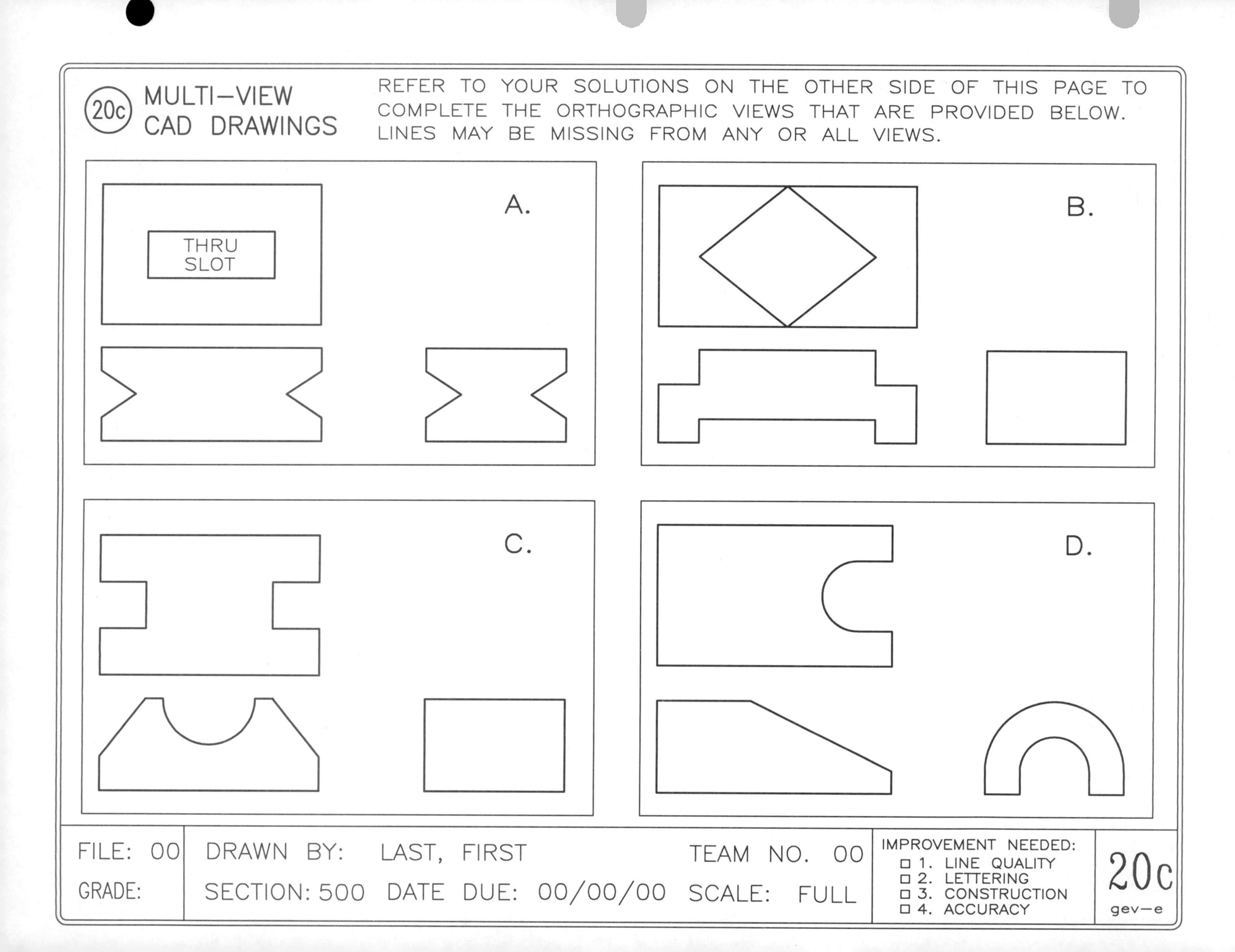
20c
MULTI–VIEW
CAD DRAWINGS
REFER TO YOUR SOLUTIONS ON THE OTHER SIDE OF THIS PAGE TO
COMPLETE THE ORTHOGRAPHIC VIEWS THAT ARE PROVIDED BELOW.
LINES MAY BE MISSING FROM ANY OR ALL VIEWS.
A.
THRU
SLOT
B.
C.
D.
FILE: 00
GRADE:
DRAWN BY: LAST, FIRST
TEAM NO. 00
SECTION: 500
DATE DUE: 00/00/00
SCALE: FULL
IMPROVEMENT NEEDED:
☐ 1. LINE QUALITY
☐ 2. LETTERING
☐ 3. CONSTRUCTION
☐ 4. ACCURACY
20c
gev–e

(21) COMPLETE THE ORTHOGRAPHIC VIEWS OF THE PARTS SHOWN BELOW. LINES MAY BE MISSING FROM ANY VIEW. SKETCH HALF SIZE PICTORIALS OF EACH PART IN THE BOX PROVIDED. THIS WILL HELP YOU VISUALIZE ANY MISSING LINES.

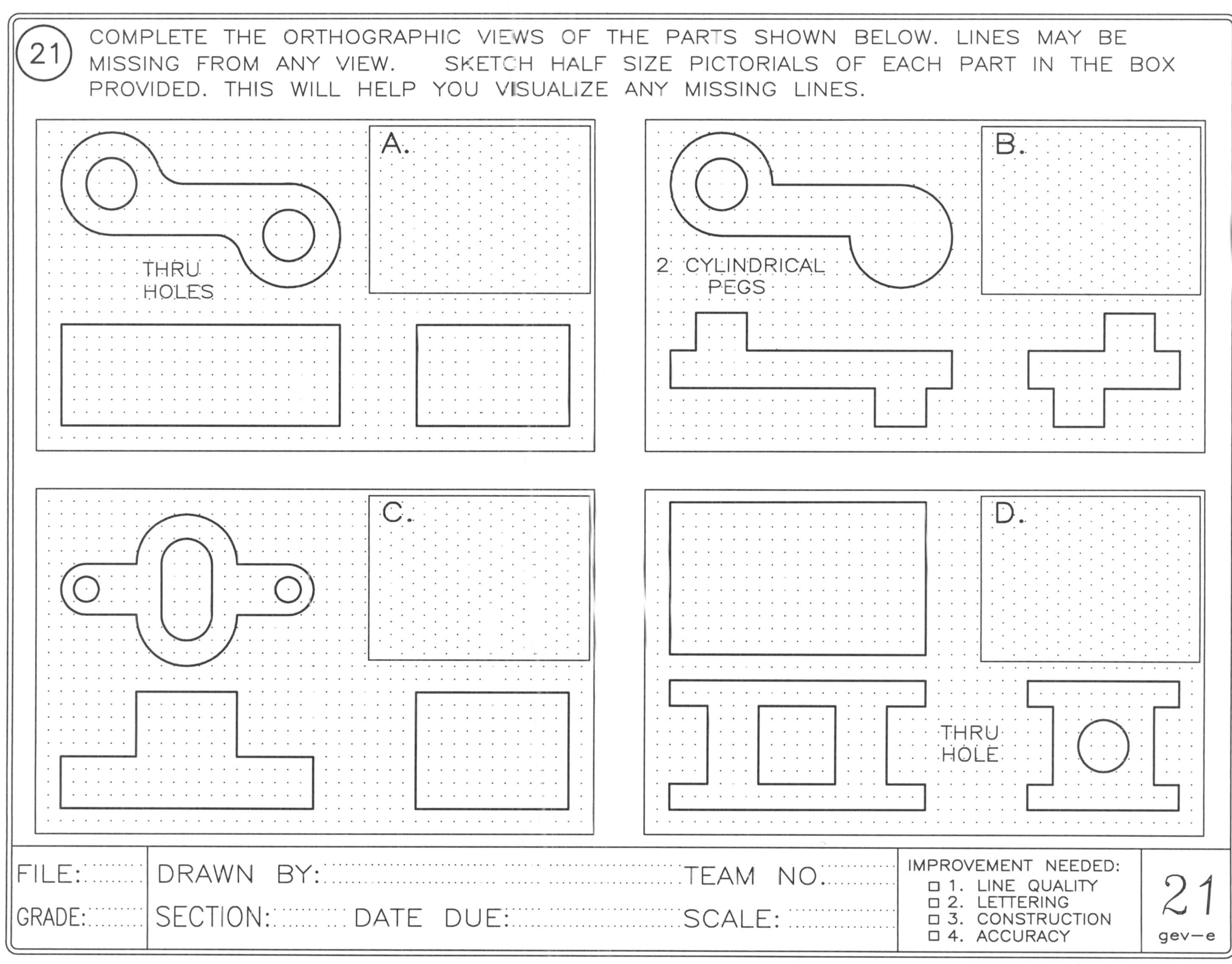

FILE:	DRAWN BY:	TEAM NO.	IMPROVEMENT NEEDED: ☐ 1. LINE QUALITY ☐ 2. LETTERING ☐ 3. CONSTRUCTION ☐ 4. ACCURACY	21 gev-e
GRADE:	SECTION: DATE DUE:	SCALE:		

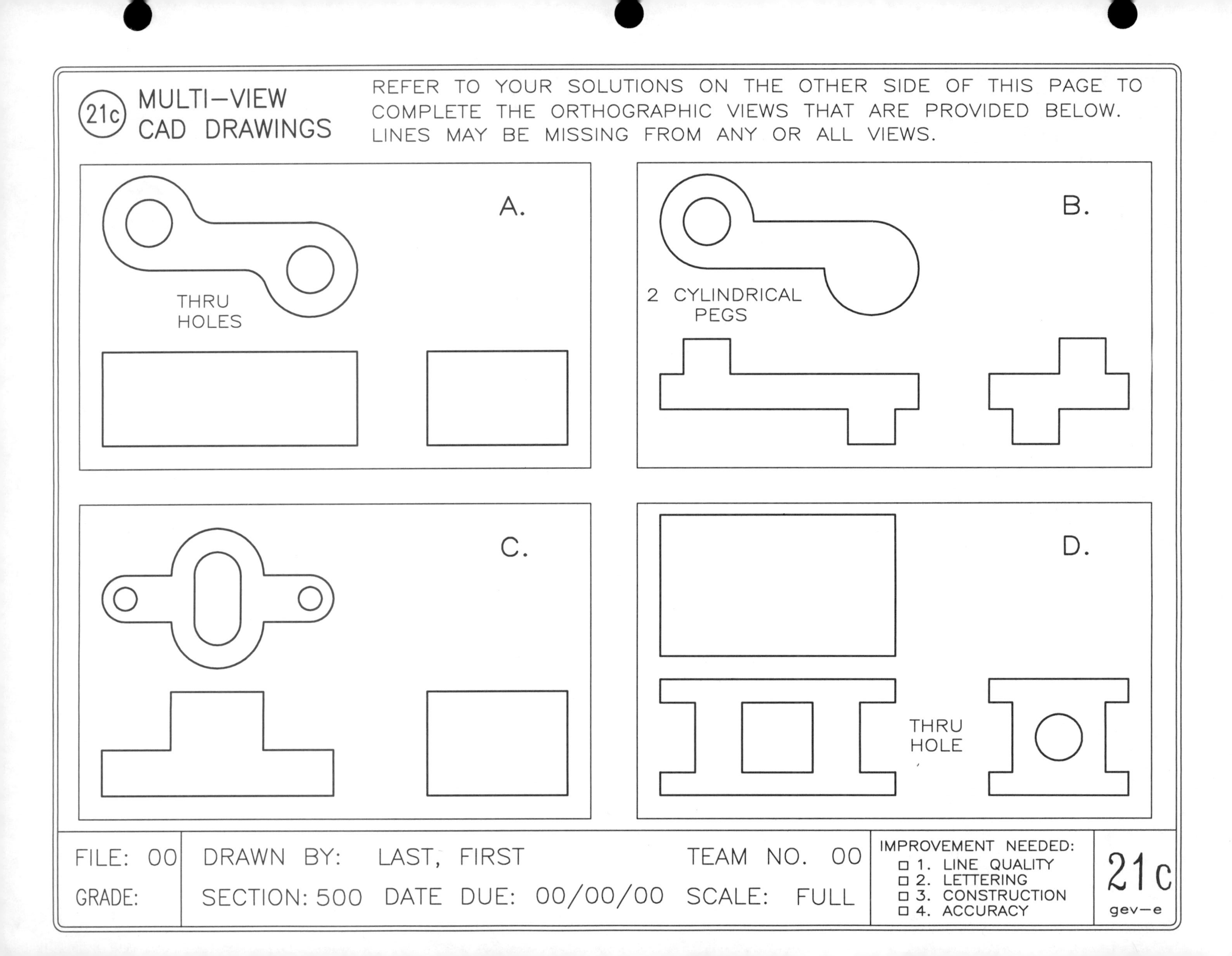
21c
MULTI–VIEW
CAD DRAWINGS
REFER TO YOUR SOLUTIONS ON THE OTHER SIDE OF THIS PAGE TO COMPLETE THE ORTHOGRAPHIC VIEWS THAT ARE PROVIDED BELOW. LINES MAY BE MISSING FROM ANY OR ALL VIEWS.
A.
THRU
HOLES
B.
2 CYLINDRICAL
PEGS
C.
D.
THRU
HOLE
FILE: 00
GRADE:
DRAWN BY: LAST, FIRST
TEAM NO. 00
SECTION: 500
DATE DUE: 00/00/00
SCALE: FULL
IMPROVEMENT NEEDED:
□ 1. LINE QUALITY
□ 2. LETTERING
□ 3. CONSTRUCTION
□ 4. ACCURACY
21c
gev–e

(22) COMPLETE THE ORTHOGRAPHIC VIEWS OF THE PARTS SHOWN BELOW. LINES MAY BE MISSING FROM ANY VIEW. SKETCH HALF SIZE PICTORIALS OF EACH PART IN THE BOX PROVIDED. THIS WILL HELP YOU VISUALIZE ANY MISSING LINES.

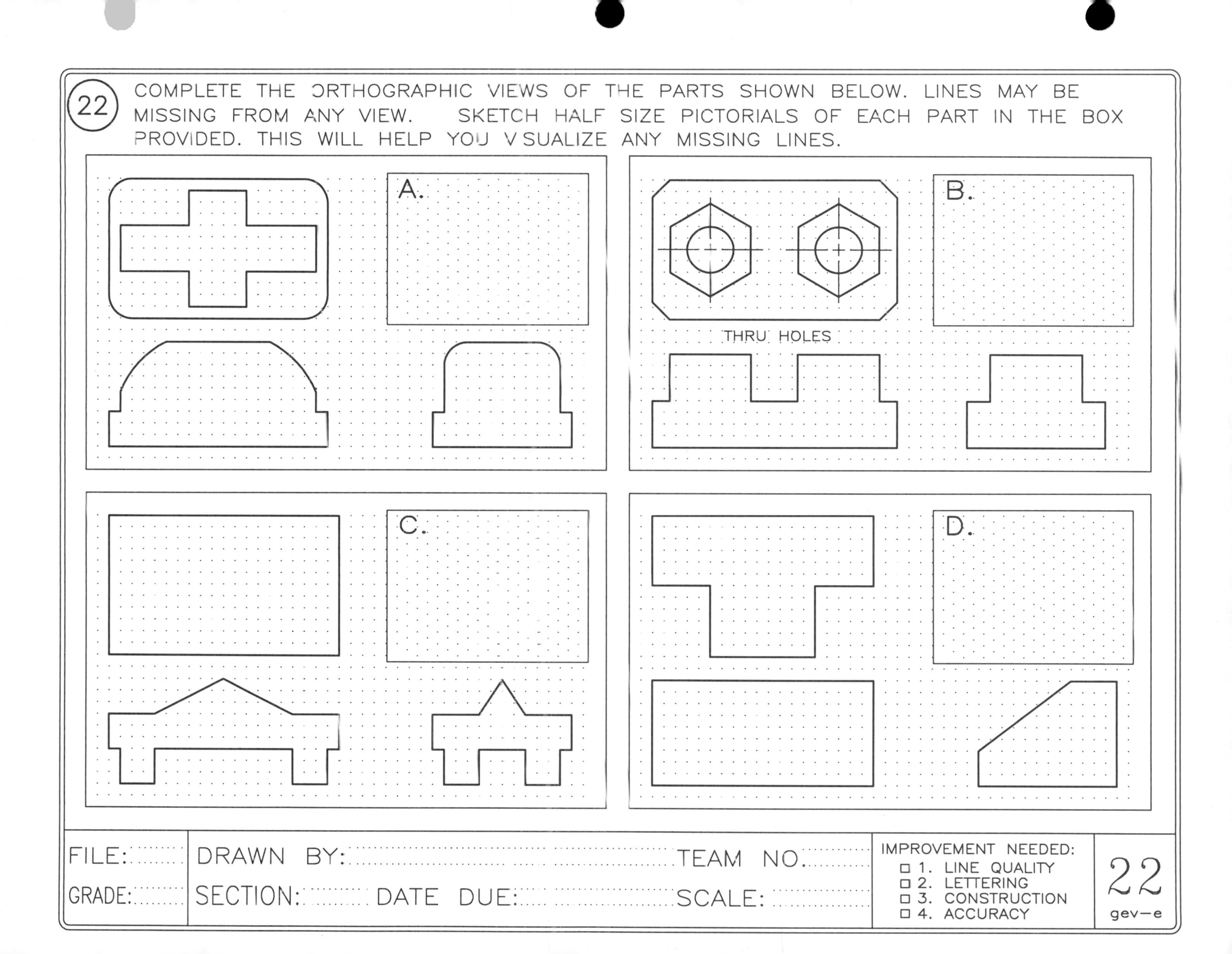

FILE:	DRAWN BY:	TEAM NO.	IMPROVEMENT NEEDED:	22
GRADE:	SECTION: DATE DUE:	SCALE:	☐ 1. LINE QUALITY ☐ 2. LETTERING ☐ 3. CONSTRUCTION ☐ 4. ACCURACY	gev—e

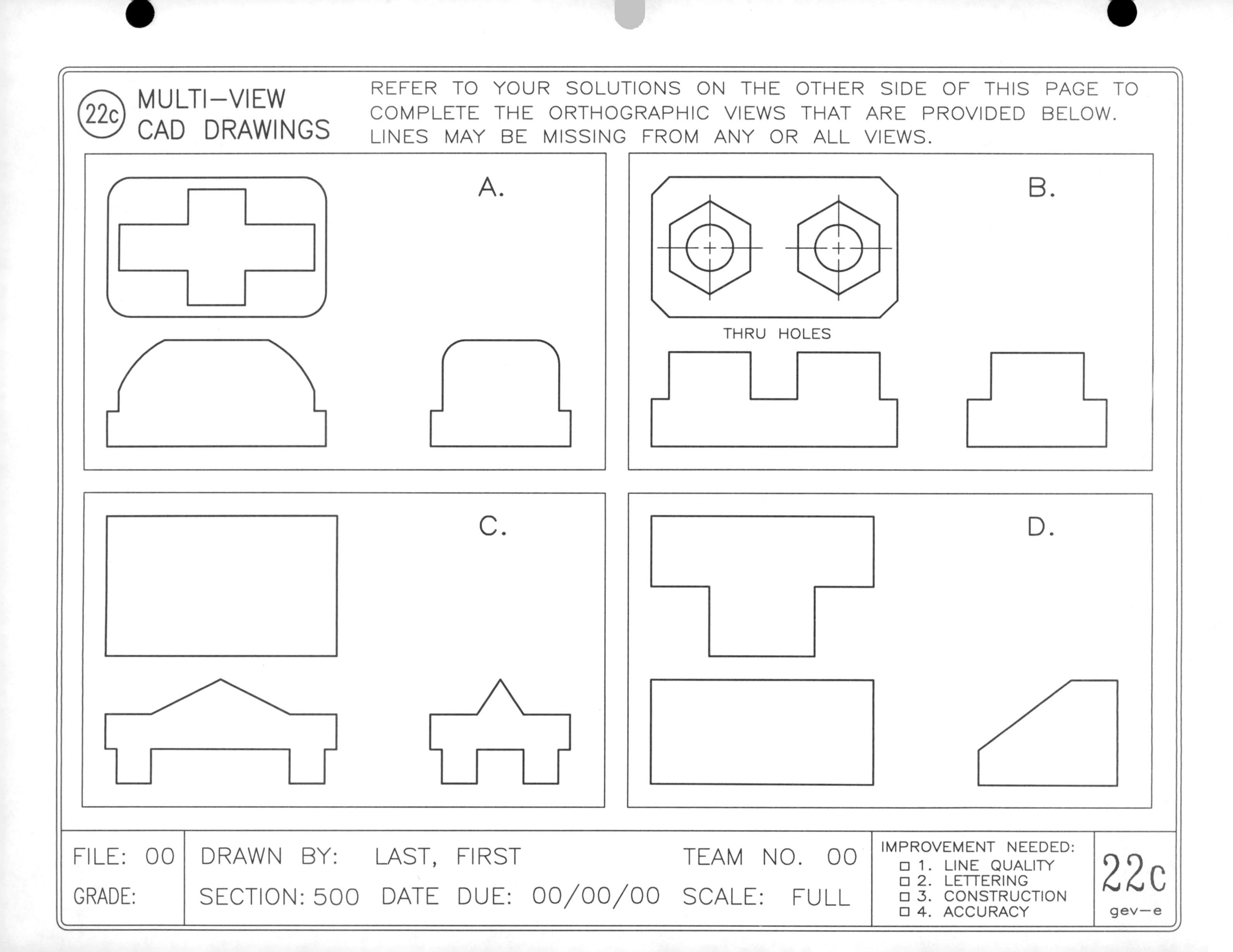
22c
MULTI–VIEW
CAD DRAWINGS
REFER TO YOUR SOLUTIONS ON THE OTHER SIDE OF THIS PAGE TO COMPLETE THE ORTHOGRAPHIC VIEWS THAT ARE PROVIDED BELOW. LINES MAY BE MISSING FROM ANY OR ALL VIEWS.
A.
B.
THRU HOLES
C.
D.
FILE: 00
GRADE:
DRAWN BY: LAST, FIRST
TEAM NO. 00
SECTION: 500 DATE DUE: 00/00/00 SCALE: FULL
IMPROVEMENT NEEDED:
☐ 1. LINE QUALITY
☐ 2. LETTERING
☐ 3. CONSTRUCTION
☐ 4. ACCURACY
22c
gev–e

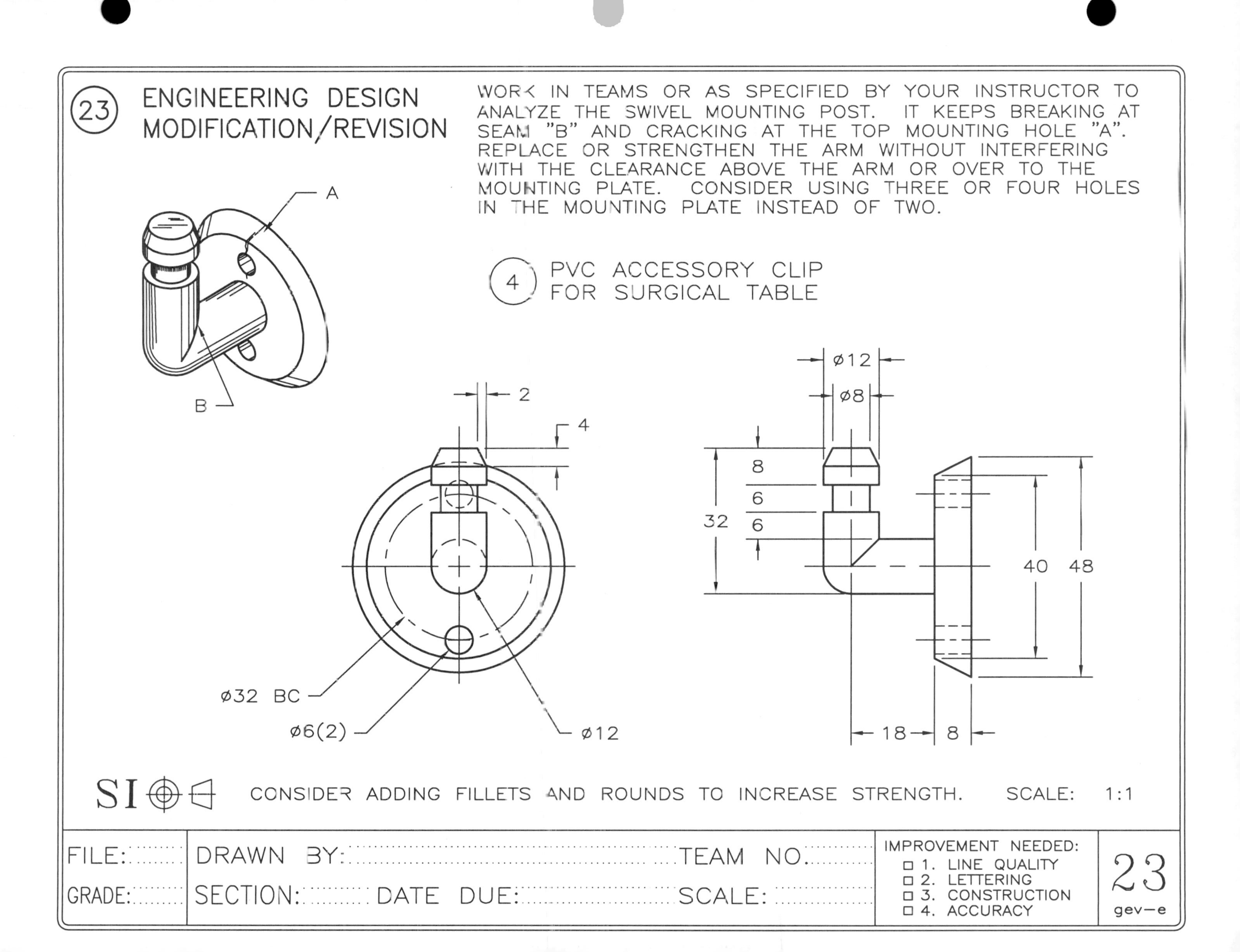

23
ENGINEERING DESIGN
MODIFICATION/REVISION
WORK IN TEAMS OR AS SPECIFIED BY YOUR INSTRUCTOR TO ANALYZE THE SWIVEL MOUNTING POST. IT KEEPS BREAKING AT SEAM "B" AND CRACKING AT THE TOP MOUNTING HOLE "A". REPLACE OR STRENGTHEN THE ARM WITHOUT INTERFERING WITH THE CLEARANCE ABOVE THE ARM OR OVER TO THE MOUNTING PLATE. CONSIDER USING THREE OR FOUR HOLES IN THE MOUNTING PLATE INSTEAD OF TWO.
A
B
4
PVC ACCESSORY CLIP
FOR SURGICAL TABLE
2
4
ø32 BC
ø6(2)
ø12
ø12
ø8
8
6
6
32
40
48
18
8
SI
CONSIDER ADDING FILLETS AND ROUNDS TO INCREASE STRENGTH.
SCALE: 1:1
FILE:
GRADE:
DRAWN BY:
TEAM NO.
SECTION:
DATE DUE:
SCALE:
IMPROVEMENT NEEDED:
1. LINE QUALITY
2. LETTERING
3. CONSTRUCTION
4. ACCURACY
23
gev-e

(23c) ENGINEERING DESIGN MODIFICATION/REVISION

(4) PVC ACCESSORY CLIP FOR SURGICAL TABLE

REFER TO YOUR SOLUTION SKETCHES FROM THE OTHER SIDE OF THIS PAGE TO CREATE ORTHOGRAPHIC CAD VIEWS OF YOUR DESIGN MODIFICATION TO THE ACCESSORY CLIP. WORK FULL SCALE, OMIT DIMENSIONS, AND ADD NOTES WHICH POINT OUT THE IMPROVEMENTS YOUR TEAM IS RECOMMENDING.

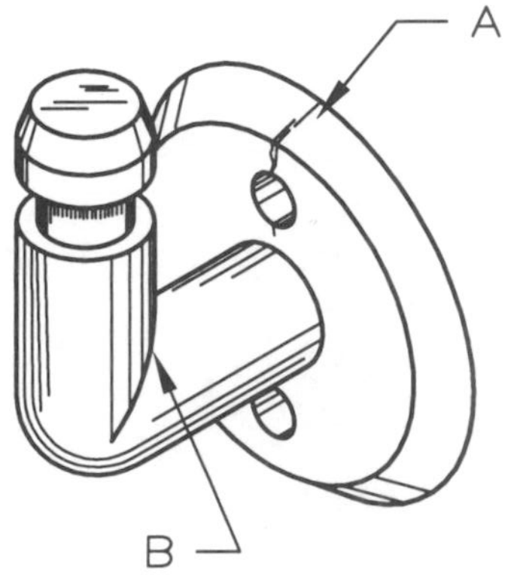

A

B

SI

FILE: 00	DRAWN BY: LAST, FIRST	TEAM NO. 00	IMPROVEMENT NEEDED:	23c
GRADE:	SECTION: 500 DATE DUE: 00/00/00	SCALE: FULL	□ 1. LINE QUALITY □ 2. LETTERING □ 3. CONSTRUCTION □ 4. ACCURACY	gev—e

(24) DESIGN ANALYSIS AND MODIFICATION SHIFTER CRANK ARM

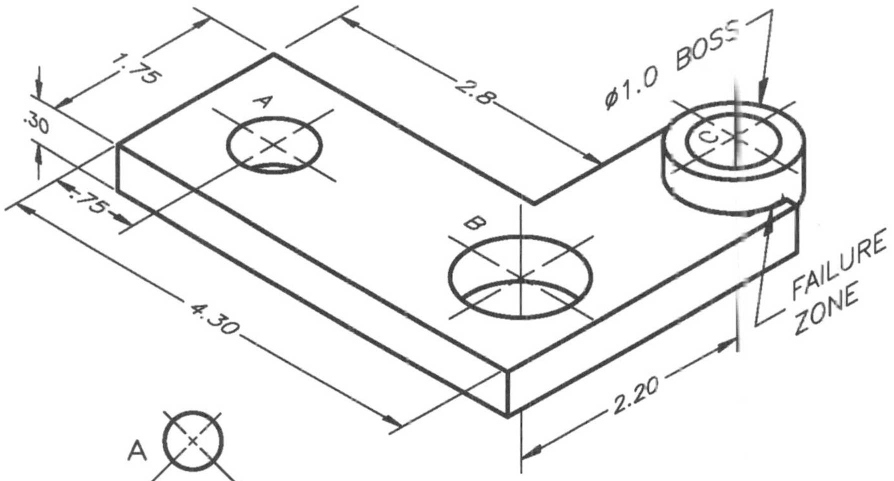

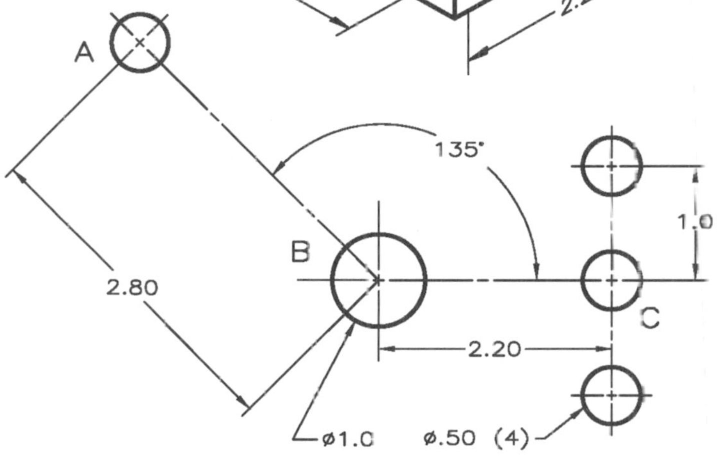

ENGINEERING CHANGE ORDER (ECO)

DESIGN SCHEMATIC FOR REVISION (NTS)

(1) SHIFTER CRANK ARM

ALUMINUM 10 REQ'D

THE DESIGN CHALLENGE

DESIGN A NEW, MORE ROBUST, SHIFTER CRANK ARM WHICH INCLUDES THESE IMPROVEMENTS:

1). ROUND OFF ALL CORNERS AND MAKE SURE THERE IS A RING OF METAL AROUND THE CIRCUMFERENCE OF EACH HOLE THAT IS AT LEAST EQUAL TO THE HOLE RADIUS.

2). INCREASE THE SEPARATION ANGLE TO 135°.

3). ADD .10" TALL BOSSES TO THE TOP AND BOTTOM OF EACH HOLE. THE BOSS DIAMETER IS APPROXIMATELY 2X THE HOLE SIZE, AND IS FILLETED WHERE IT JOINS THE BASE.

4) INCREASE THE RANGE OF ADJUSTMENT AND ADAPTABILITY BY ADDING TWO ADDITIONAL HOLES AS INDICATED ON EITHER SIDE OF HOLE "C".

AS INDIVIDUALS, OR IN TEAMS ASSIGNED BY YOUR INSTRUCTOR, PROVIDE ORTHOGRAPHIC DRAWINGS OF YOUR DESIGN SOLUTION INCLUDING PRELIMINARY SKETCHES OF VARIOUS DESIGN POSSIBILITIES. USE WORK SHEETS FROM THE BACK OF THIS BOOK. WORK FULL SCALE AND LEAVE OFF DIMENSIONS. ESTABLISH ANY MISSING DIMENSIONS REQUIRED FOR YOUR DESIGN.

FILE:	DRAWN BY:	TEAM NO.	IMPROVEMENT NEEDED: ☐ 1. LINE QUALITY ☐ 2. LETTERING ☐ 3. CONSTRUCTION ☐ 4. ACCURACY	24
GRADE:	SECTION: DATE DUE:	SCALE:		gev–e

(24c) DESIGN MODIFICATION

REFER TO YOUR SOLUTIOINS FROM THE OTHER SIDE OF THIS PAGE AND MAKE A FULL SIZE CAD DRAWING OF THE TOP VIEW.

(1) SHIFTER CRANK ARM

ALUMINUM 10 REQ'D

FILE: 00	DRAWN BY: LAST, FIRST	TEAM NO. 00	IMPROVEMENT NEEDED:	24c
GRADE:	SECTION: 500 DATE DUE: 00/00/00	SCALE: FULL	☐ 1. LINE WEIGHTS ☐ 2. LAYOUT ☐ 3. CONSTRUCTION ☐ 4. ACCURACY	gev—e

(25) LETTERING MASTERY

COPY EACH SENTENCE ONCE OR TWICE AS ASSIGNED BY YOUR INSTRUCTOR. PRACTICE PROPER SPACING BETWEEN LETTERS, LINES AND WORDS. USE A SOFT, MEDIUM POINT LEAD AND ALLOW A FULL SPACE BETWEEN EACH SENTENCE.

1. ONE OF THE THREE AXONOMETRIC PICTORIALS IS CALLED ISOMETRIC.

2. AXONOMETRIC, OBLIQUE, OR PERSPECTIVE DRAWINGS ARE UTILIZED.

3. CAVALIER OBLIQUES ARE ALWAYS DRAWN WITH A FULL SIZE DEPTH.

4. CABINET OBLIQUES ARE ALWAYS DRAWN WITH A HALF SIZE DEPTH.

5. OBLIQUE PICTORIALS MAY HAVE BASE ANGLES FROM 0° TO 90°.

6. ISOMETRIC PICTORIALS MUST ALWAYS HAVE THEIR AXES 120° APART.

7. MOST CIRCLES BECOME 35° ELLIPSES WHEN PLOTTED AS ISOMETRICS.

FILE: .... DRAWN BY: .... TEAM NO. ....

GRADE: .... SECTION: .... DATE DUE: .... SCALE: ....

IMPROVEMENT NEEDED:
- ☐ 1. LINE QUALITY
- ☐ 2. LETTERING
- ☐ 3. CONSTRUCTION
- ☐ 4. ACCURACY

25

gev–e

25c LETTERING MASTERY

TYPE EACH MEMORY LINE FROM THE OTHER SIDE OF THIS PAGE ONCE OR TWICE AS ASSIGNED BY YOUR INSTRUCTOR. USE A DIFFERENT TEXT STYLE AND FONT FOR EACH OF THE SEVEN SENTENCES.

1.

2.

3.

4.

5.

6.

7.

FILE: 00	DRAWN BY: LAST, FIRST	TEAM NO. 00	IMPROVEMENT NEEDED: ☐ 1. LINE QUALITY ☐ 2. LETTERING ☐ 3. CONSTRUCTION ☐ 4. ACCURACY	25c gev–e
GRADE:	SECTION: 500 DATE DUE: 00/00/00	SCALE: FULL		

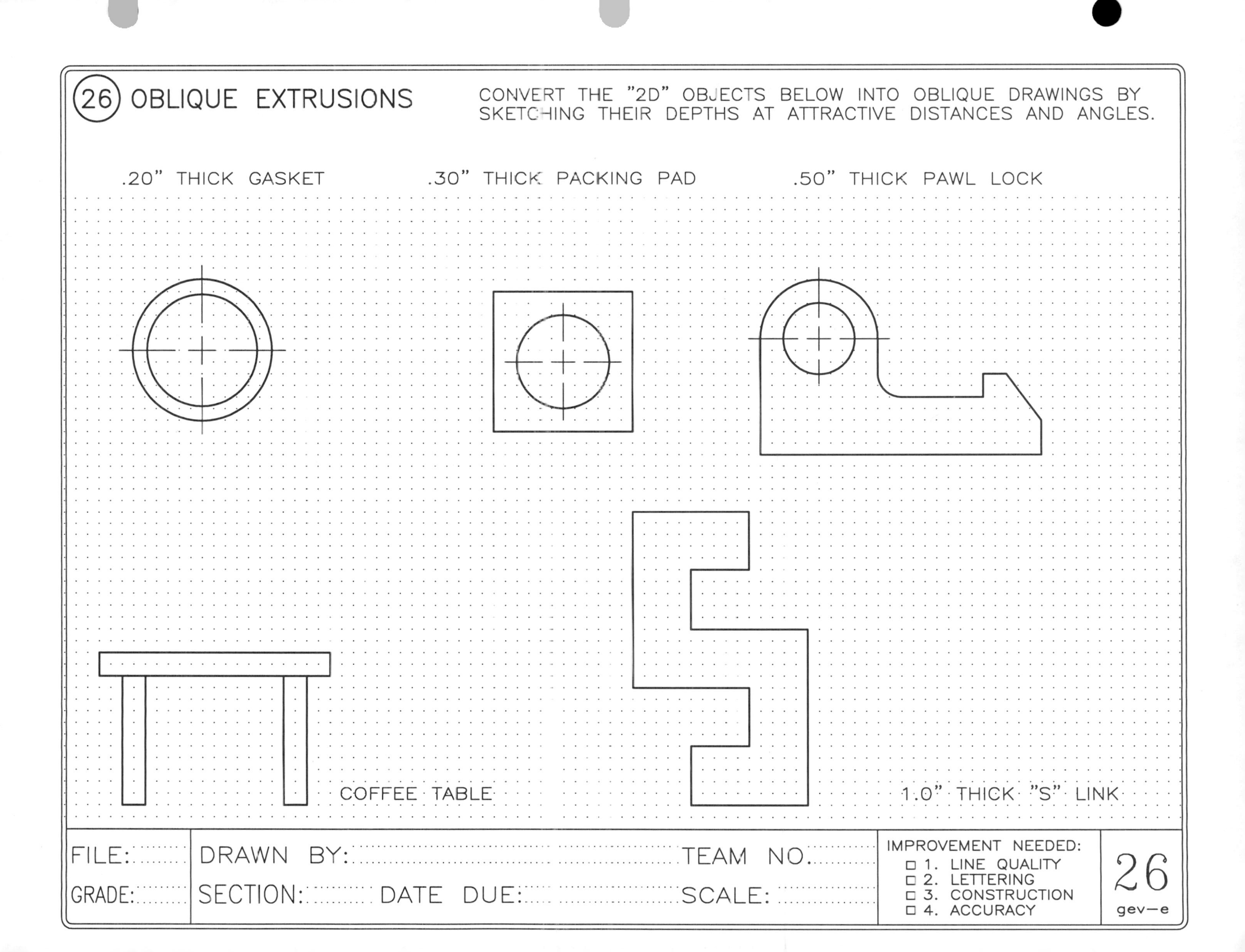

26 OBLIQUE EXTRUSIONS
CONVERT THE "2D" OBJECTS BELOW INTO OBLIQUE DRAWINGS BY SKETCHING THEIR DEPTHS AT ATTRACTIVE DISTANCES AND ANGLES.
.20" THICK GASKET
.30" THICK PACKING PAD
.50" THICK PAWL LOCK
COFFEE TABLE
1.0" THICK "S" LINK
FILE:
GRADE:
DRAWN BY:
TEAM NO.
SECTION:
DATE DUE:
SCALE:
IMPROVEMENT NEEDED:
☐ 1. LINE QUALITY
☐ 2. LETTERING
☐ 3. CONSTRUCTION
☐ 4. ACCURACY
26
gev–e

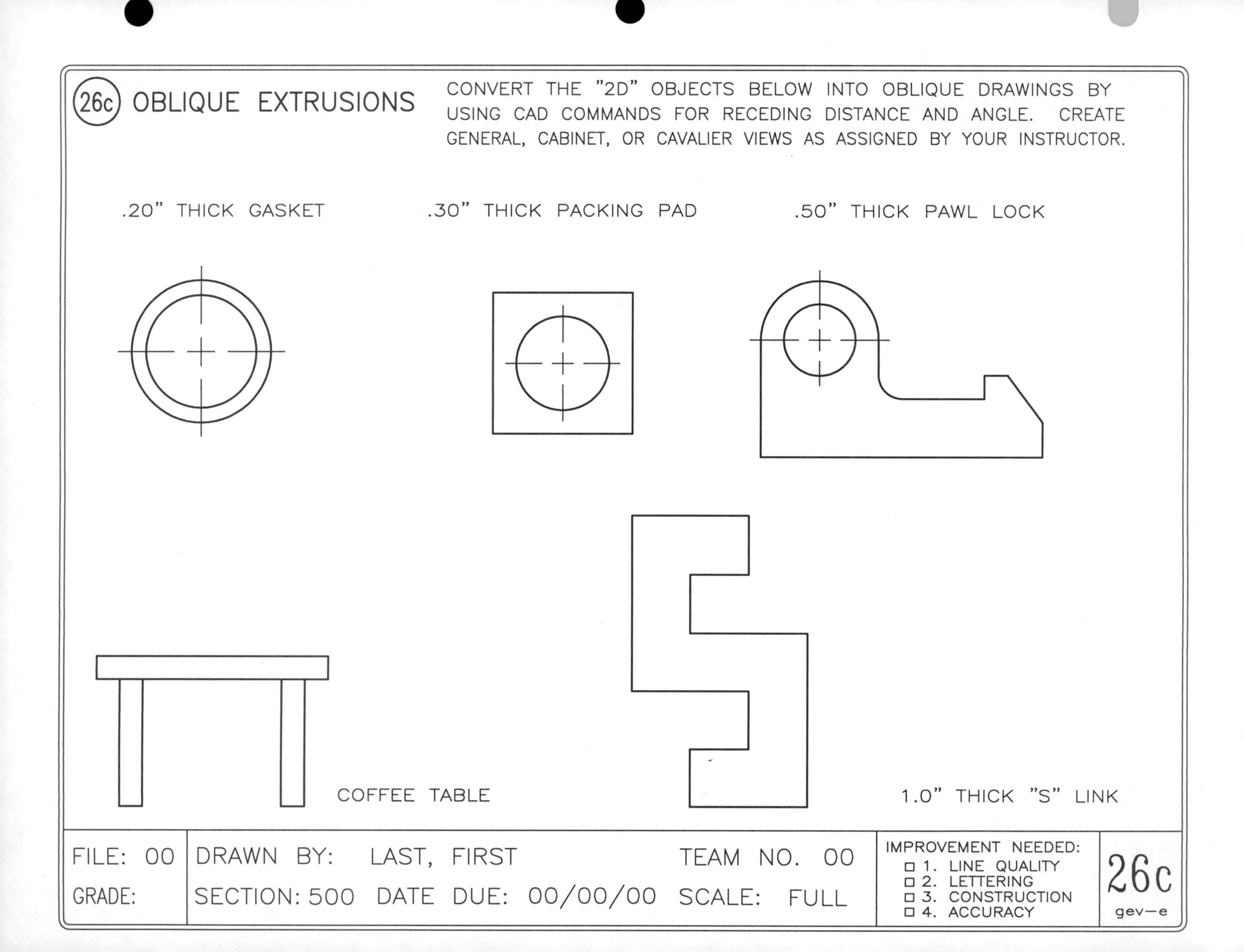
26c
OBLIQUE EXTRUSIONS
CONVERT THE "2D" OBJECTS BELOW INTO OBLIQUE DRAWINGS BY USING CAD COMMANDS FOR RECEDING DISTANCE AND ANGLE. CREATE GENERAL, CABINET, OR CAVALIER VIEWS AS ASSIGNED BY YOUR INSTRUCTOR.
.20" THICK GASKET
.30" THICK PACKING PAD
.50" THICK PAWL LOCK
COFFEE TABLE
1.0" THICK "S" LINK
FILE: 00
GRADE:
DRAWN BY: LAST, FIRST
TEAM NO. 00
SECTION: 500
DATE DUE: 00/00/00
SCALE: FULL
IMPROVEMENT NEEDED:
1. LINE QUALITY
2. LETTERING
3. CONSTRUCTION
4. ACCURACY
26c
gev-e

(28) OBLIQUE PICTORIALS

SKETCH CABINET OR CAVALIER OBLIQUES OF THE OBJECTS SHOWN BELOW AS ASSIGNED BY YOUR INSTRUCTOR. USE THE ANGLE INDICATED BY THE "GRID-BOX" AND SELECT THE FRONT VIEW BEST SUITED FOR OBLIQUE DRAWINGS.

A.

B.

C.

MAKE YOUR SKETCHES DOUBLE SIZE (2:1)

TYPE: ....

TYPE: ....

TYPE: ....

FILE: ....
GRADE: ....
DRAWN BY: .... TEAM NO. ....
SECTION: .... DATE DUE: .... SCALE: ....

IMPROVEMENT NEEDED:
- ☐ 1. LINE QUALITY
- ☐ 2. LETTERING
- ☐ 3. CONSTRUCTION
- ☐ 4. ACCURACY

gev-e

(28c) OBLIQUE PICTORIALS

DRAW DOUBLE SIZE CABINET OR CAVALIER OBLIQUES OF THE OBJECTS SHOWN BELOW AS SPECIFIED BY YOUR INSTRUCTOR. USE ATTRACTIVE BASE ANGLES OF YOUR CHOICE AND SELECT THE FRONT VIEW BEST SUITED FOR EACH OBJECT.

A.

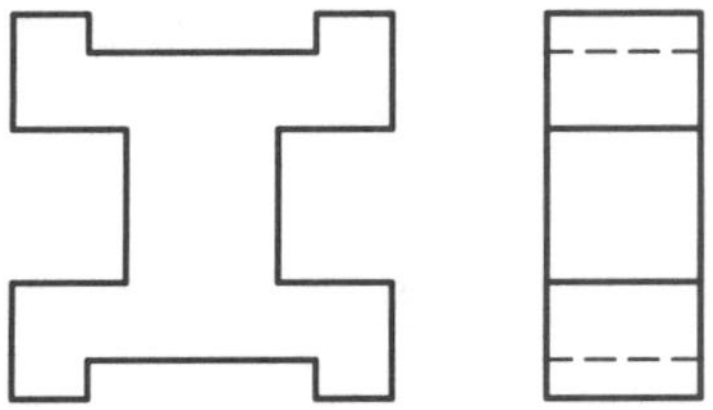

B.

C.

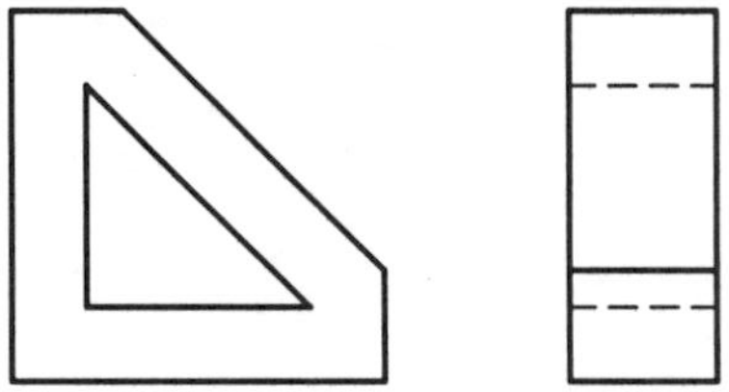

MAKE YOUR DRAWINGS DOUBLE SIZE (2:1)

L
TYPE:

L
TYPE:

L
TYPE:

FILE: 00	DRAWN BY: LAST, FIRST	TEAM NO. 00	IMPROVEMENT NEEDED:	28c
GRADE:	SECTION: 500 DATE DUE: 00/00/00	SCALE: FULL	☐ 1. LINE QUALITY ☐ 2. LETTERING ☐ 3. CONSTRUCTION ☐ 4. ACCURACY	gev-e

(29) OBLIQUE PICTORIALS

SKETCH CABINET OR CAVALIER OBLIQUES OF THE OBJECTS SHOWN BELOW AS ASSIGNED BY YOUR INSTRUCTOR. USE THE ANGLE INDICATED BY THE "GRID-BOX" AND SELECT THE FRONT VIEW BEST SUITED FOR OBLIQUE DRAWINGS.

A.

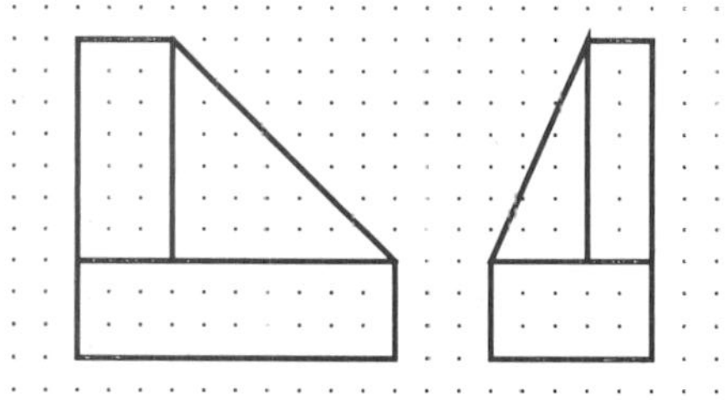

B.

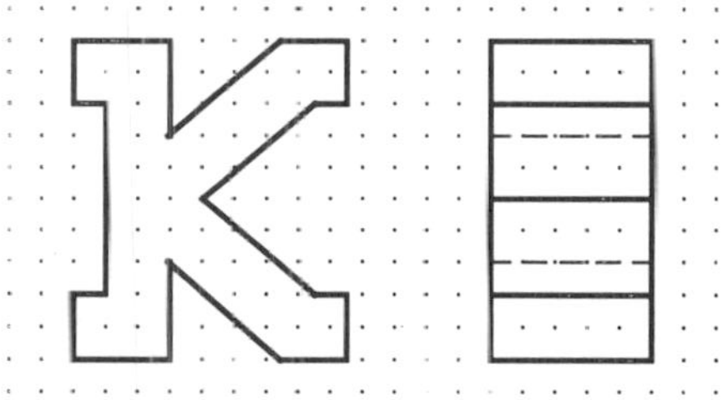

C.

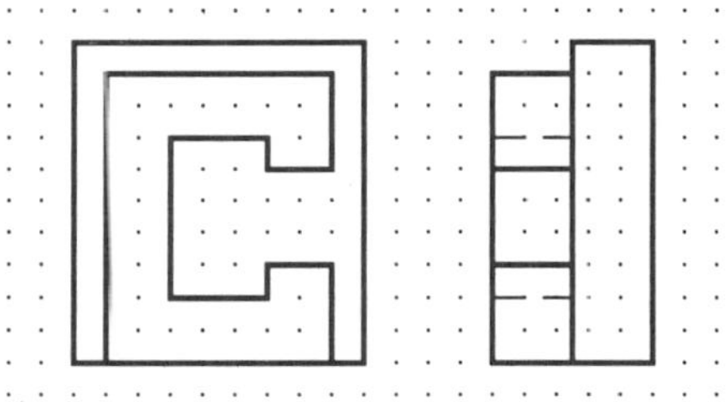

MAKE YOUR SKETCHES DOUBLE SIZE (2:1)

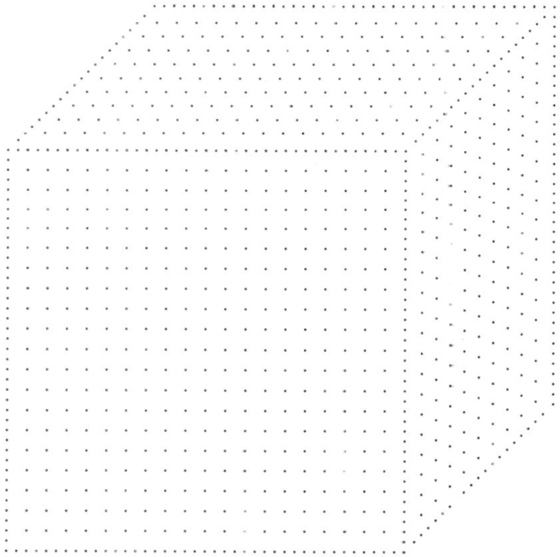

TYPE:

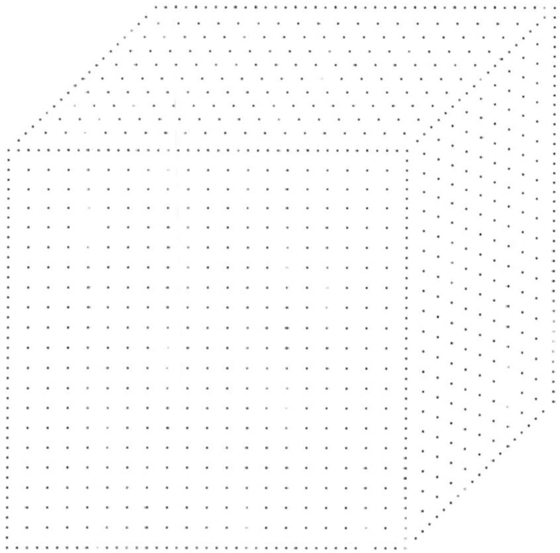

TYPE:

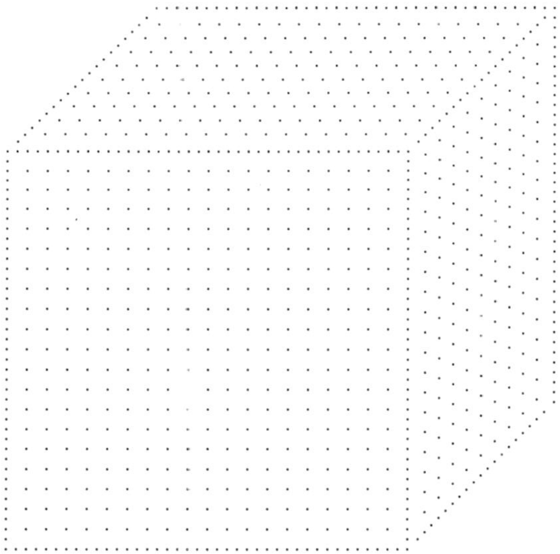

TYPE:

FILE:	DRAWN BY:	TEAM NO.	IMPROVEMENT NEEDED:	29
GRADE:	SECTION: DATE DUE:	SCALE:	☐ 1. LINE QUALITY ☐ 2. LETTERING ☐ 3. CONSTRUCTION ☐ 4. ACCURACY	gev–e

29c OBLIQUE PICTORIALS

DRAW DOUBLE SIZE CABINET OR CAVALIER OBLIQUES OF THE OBJECTS SHOWN BELOW AS SPECIFIED BY YOUR INSTRUCTOR. USE ATTRACTIVE BASE ANGLES OF YOUR CHOICE AND SELECT THE FRONT VIEW BEST SUITED FOR EACH OBJECT.

A.

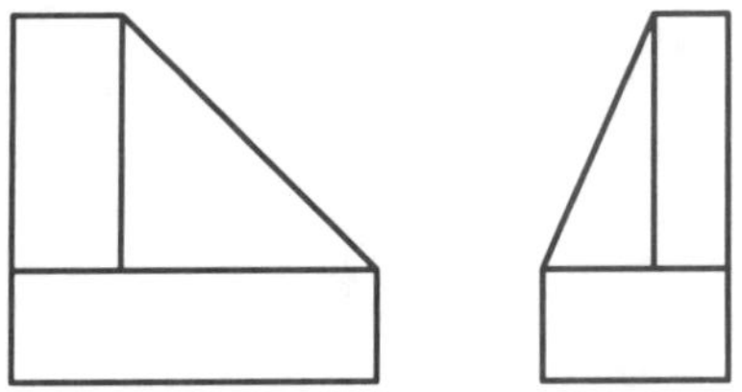

B.

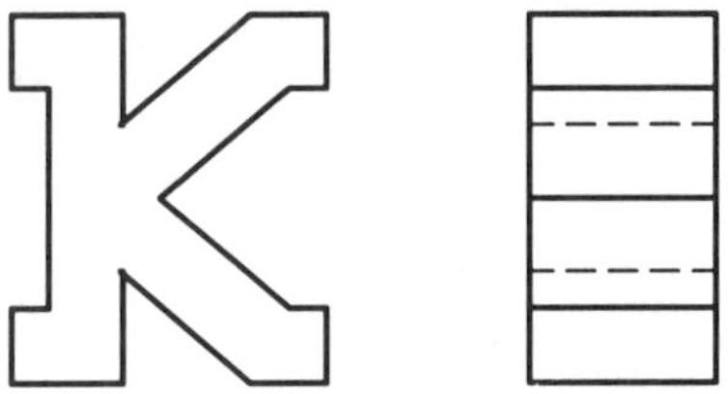

C.

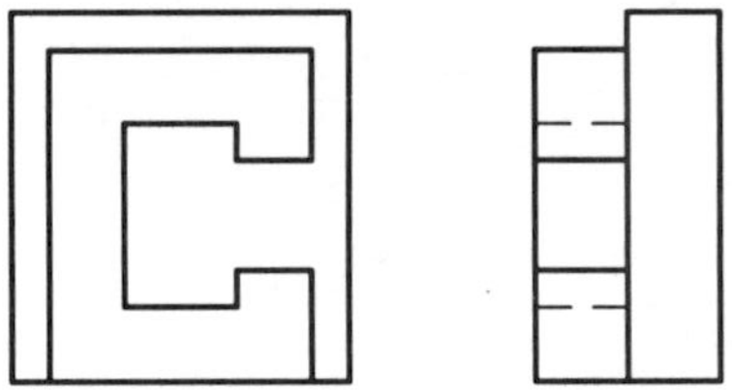

MAKE YOUR DRAWINGS DOUBLE SIZE (2:1)

∟ TYPE:

∟ TYPE:

∟ TYPE:

FILE: 00	DRAWN BY: LAST, FIRST	TEAM NO. 00	IMPROVEMENT NEEDED:	29c
GRADE:	SECTION: 500 DATE DUE: 00/00/00	SCALE: FULL	☐ 1. LINE QUALITY ☐ 2. LETTERING ☐ 3. CONSTRUCTION ☐ 4. ACCURACY	gev–e

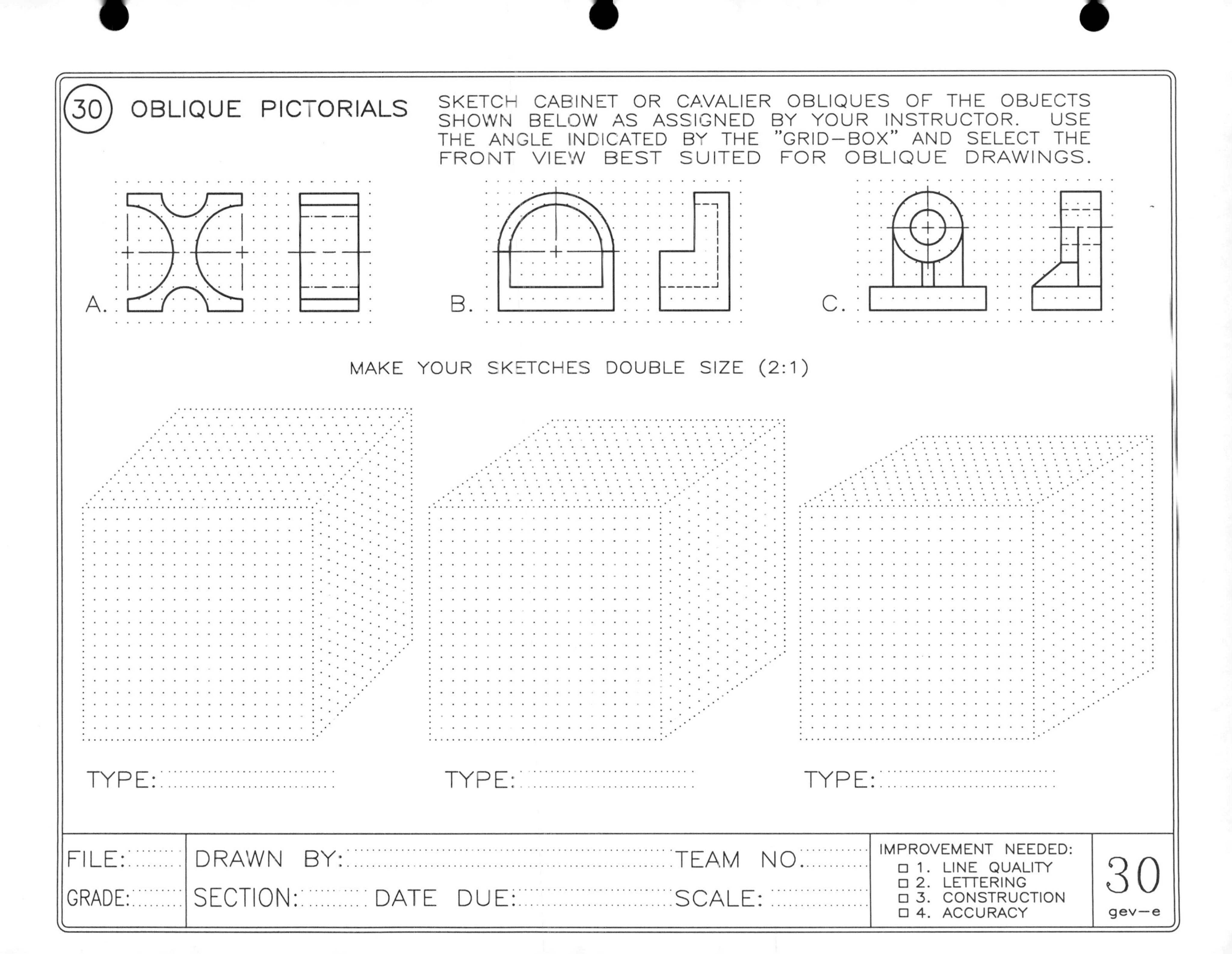
30
OBLIQUE PICTORIALS
SKETCH CABINET OR CAVALIER OBLIQUES OF THE OBJECTS SHOWN BELOW AS ASSIGNED BY YOUR INSTRUCTOR. USE THE ANGLE INDICATED BY THE "GRID-BOX" AND SELECT THE FRONT VIEW BEST SUITED FOR OBLIQUE DRAWINGS.
A.
B.
C.
MAKE YOUR SKETCHES DOUBLE SIZE (2:1)
TYPE:
TYPE:
TYPE:
FILE:
GRADE:
DRAWN BY:
TEAM NO.
SECTION:
DATE DUE:
SCALE:
IMPROVEMENT NEEDED:
☐ 1. LINE QUALITY
☐ 2. LETTERING
☐ 3. CONSTRUCTION
☐ 4. ACCURACY
30
gev-e

(30c) # OBLIQUE PICTORIALS

DRAW DOUBLE SIZE CABINET OR CAVALIER OBLIQUES OF THE OBJECTS SHOWN BELOW AS SPECIFIED BY YOUR INSTRUCTOR. USE ATTRACTIVE BASE ANGLES OF YOUR CHOICE AND SELECT THE FRONT VIEW BEST SUITED FOR EACH OBJECT.

A.

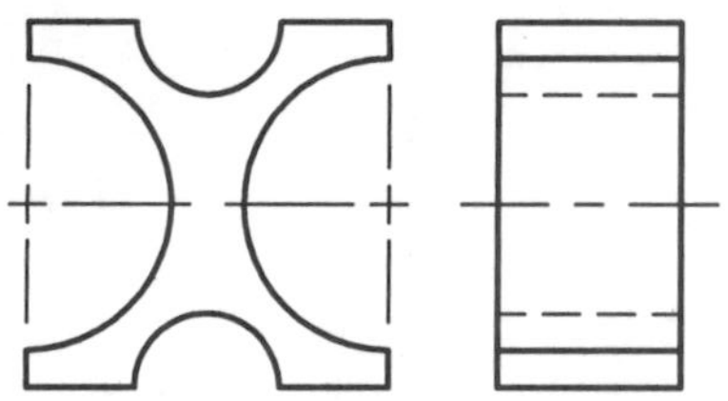

B.

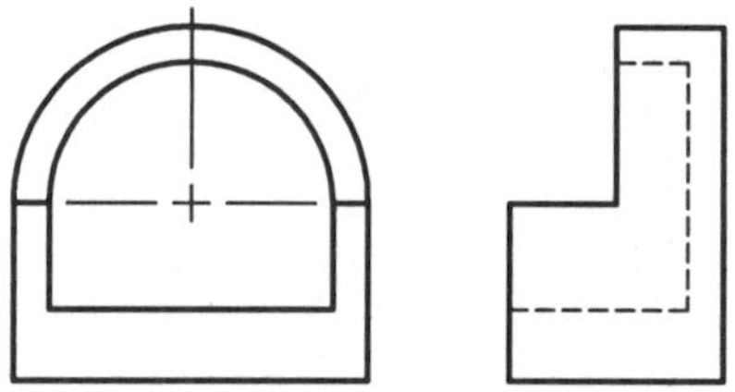

C.

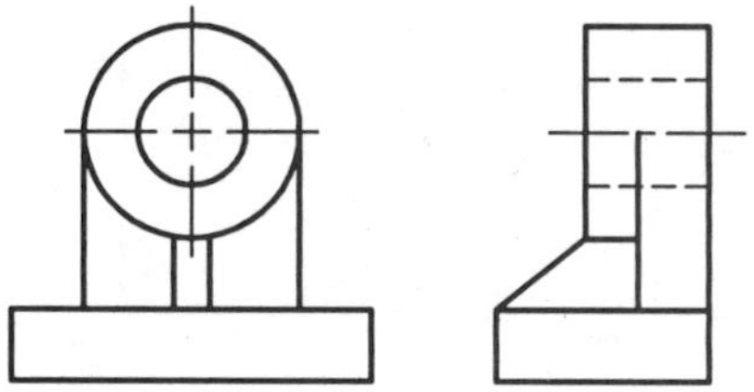

MAKE YOUR DRAWINGS DOUBLE SIZE (2:1)

∟ TYPE:

∟ TYPE:

∟ TYPE:

FILE: 00	DRAWN BY: LAST, FIRST	TEAM NO. 00	IMPROVEMENT NEEDED:	30c
GRADE:	SECTION: 500 DATE DUE: 00/00/00	SCALE: FULL	□ 1. LINE QUALITY □ 2. LETTERING □ 3. CONSTRUCTION □ 4. ACCURACY	gev–e

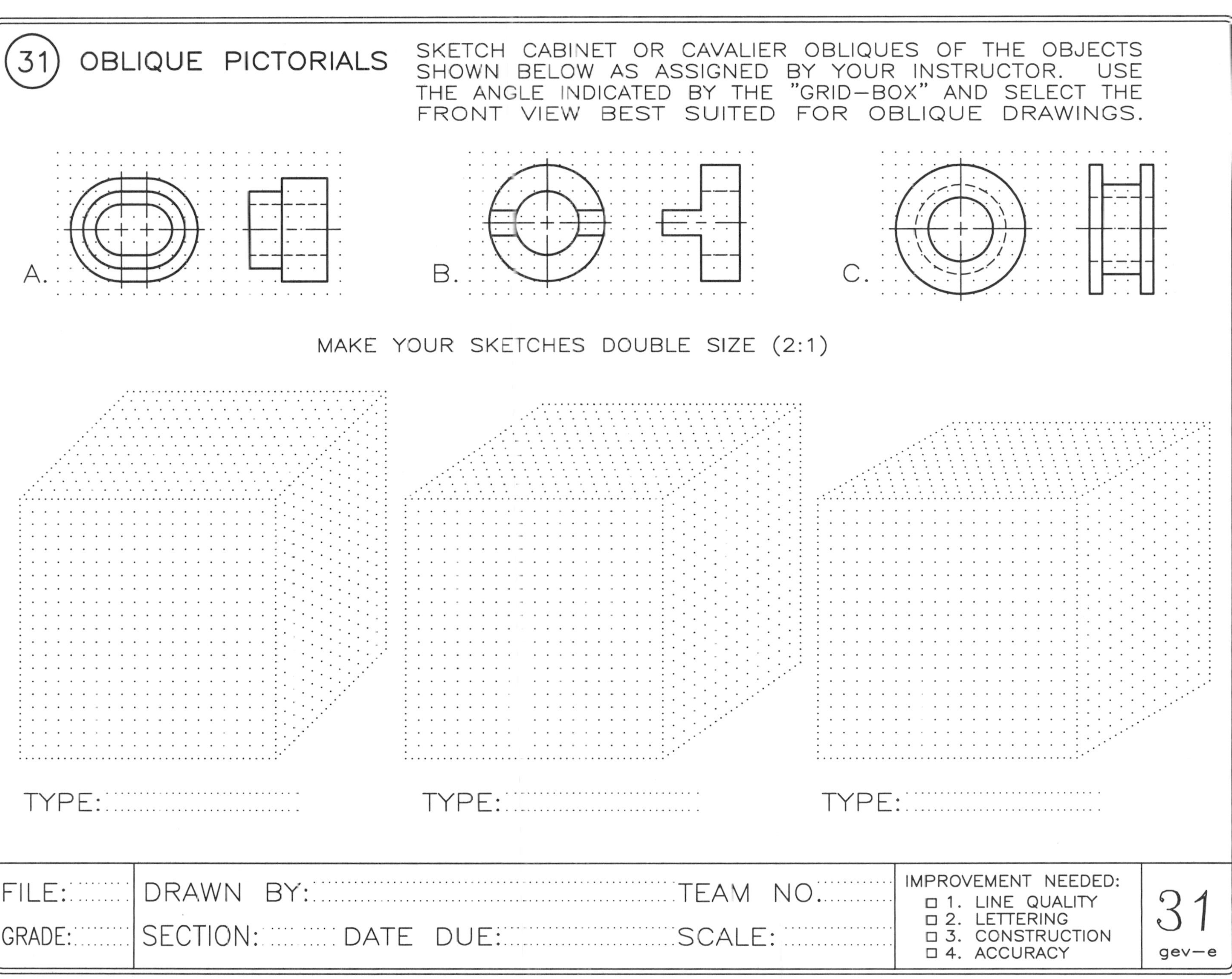
31
OBLIQUE PICTORIALS
SKETCH CABINET OR CAVALIER OBLIQUES OF THE OBJECTS SHOWN BELOW AS ASSIGNED BY YOUR INSTRUCTOR. USE THE ANGLE INDICATED BY THE "GRID-BOX" AND SELECT THE FRONT VIEW BEST SUITED FOR OBLIQUE DRAWINGS.
A.
B.
C.
MAKE YOUR SKETCHES DOUBLE SIZE (2:1)
TYPE:
TYPE:
TYPE:
FILE:
GRADE:
DRAWN BY:
TEAM NO.
SECTION:
DATE DUE:
SCALE:
IMPROVEMENT NEEDED:
☐ 1. LINE QUALITY
☐ 2. LETTERING
☐ 3. CONSTRUCTION
☐ 4. ACCURACY
31
gev-e

(31c) OBLIQUE PICTORIALS

DRAW DOUBLE SIZE CABINET OR CAVALIER OBLIQUES OF THE OBJECTS SHOWN BELOW AS SPECIFIED BY YOUR INSTRUCTOR. USE ATTRACTIVE BASE ANGLES OF YOUR CHOICE AND SELECT THE FRONT VIEW BEST SUITED FOR EACH OBJECT.

A.

B.

C.

MAKE YOUR DRAWINGS DOUBLE SIZE (2:1)

∟ TYPE:

∟ TYPE:

∟ TYPE:

FILE: 00	DRAWN BY: LAST, FIRST	TEAM NO. 00	IMPROVEMENT NEEDED:	31c
GRADE:	SECTION: 500 DATE DUE: 00/00/00	SCALE: FULL	☐ 1. LINE QUALITY ☐ 2. LETTERING ☐ 3. CONSTRUCTION ☐ 4. ACCURACY	gev–e

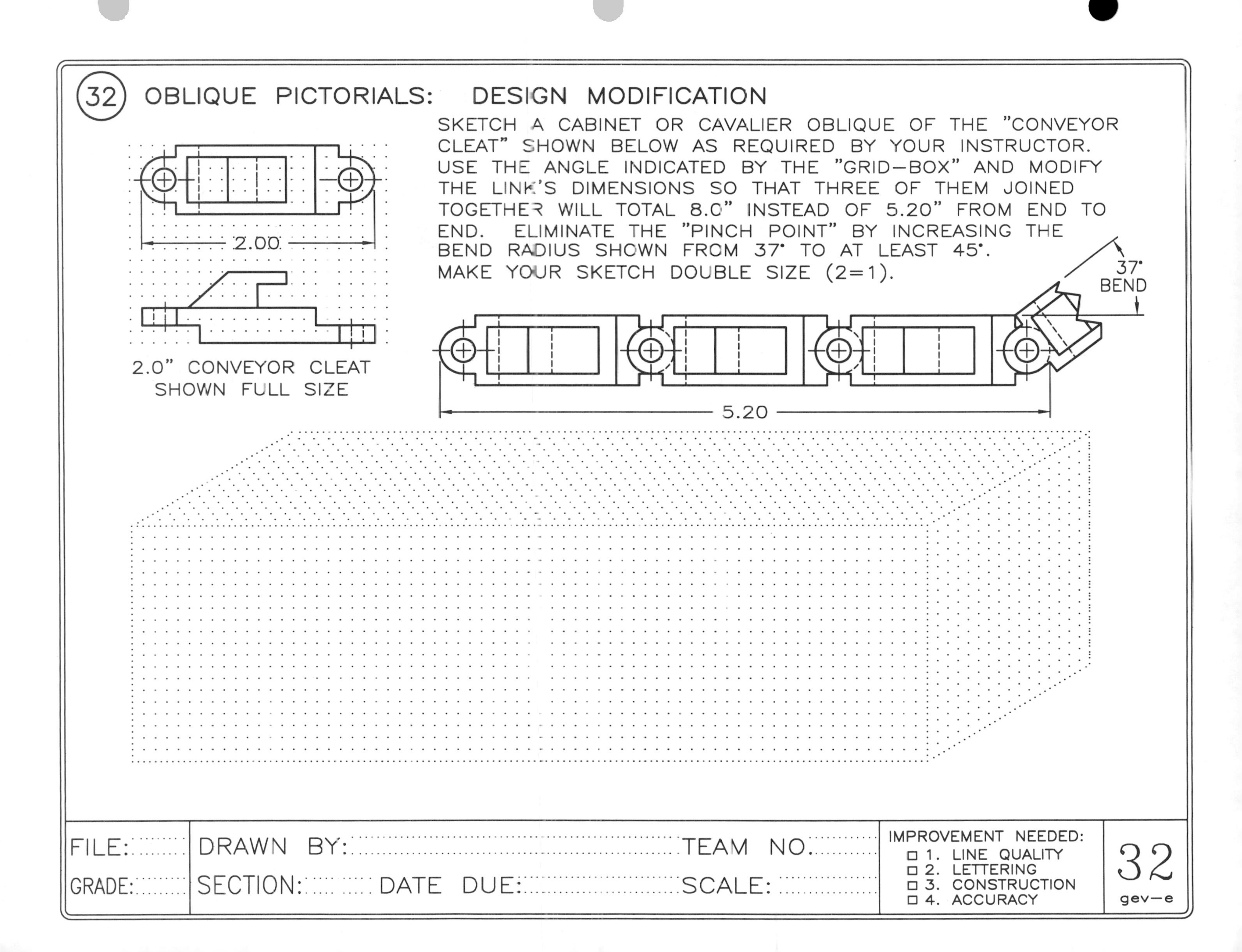
32
OBLIQUE PICTORIALS: DESIGN MODIFICATION
SKETCH A CABINET OR CAVALIER OBLIQUE OF THE "CONVEYOR CLEAT" SHOWN BELOW AS REQUIRED BY YOUR INSTRUCTOR. USE THE ANGLE INDICATED BY THE "GRID-BOX" AND MODIFY THE LINK'S DIMENSIONS SO THAT THREE OF THEM JOINED TOGETHER WILL TOTAL 8.0" INSTEAD OF 5.20" FROM END TO END. ELIMINATE THE "PINCH POINT" BY INCREASING THE BEND RADIUS SHOWN FROM 37° TO AT LEAST 45°. MAKE YOUR SKETCH DOUBLE SIZE (2=1).
2.00
2.0" CONVEYOR CLEAT SHOWN FULL SIZE
37° BEND
5.20
FILE:
GRADE:
DRAWN BY:
TEAM NO.
SECTION:
DATE DUE:
SCALE:
IMPROVEMENT NEEDED:
□ 1. LINE QUALITY
□ 2. LETTERING
□ 3. CONSTRUCTION
□ 4. ACCURACY
32
gev-e

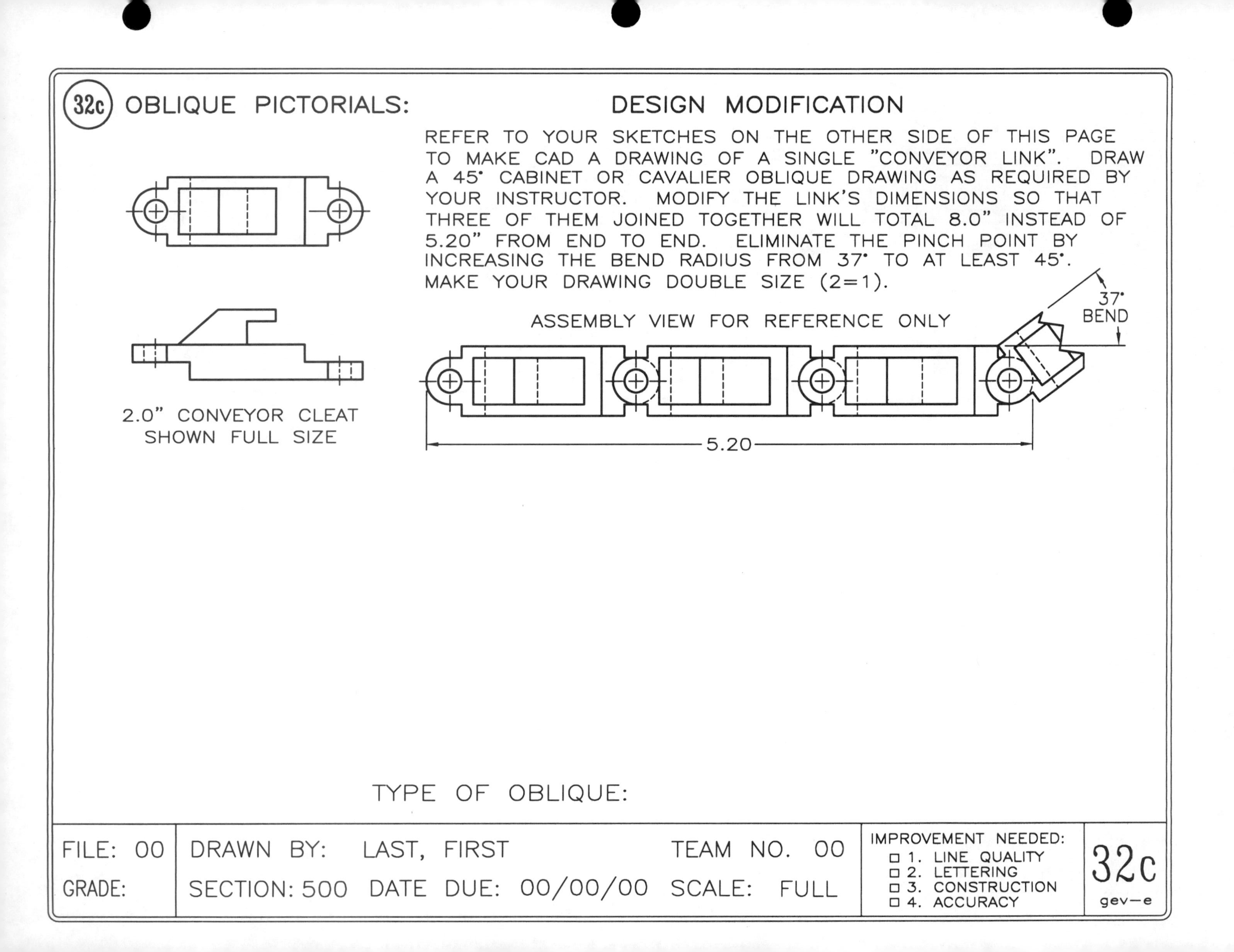
32c
OBLIQUE PICTORIALS:
DESIGN MODIFICATION
REFER TO YOUR SKETCHES ON THE OTHER SIDE OF THIS PAGE TO MAKE CAD A DRAWING OF A SINGLE "CONVEYOR LINK". DRAW A 45° CABINET OR CAVALIER OBLIQUE DRAWING AS REQUIRED BY YOUR INSTRUCTOR. MODIFY THE LINK'S DIMENSIONS SO THAT THREE OF THEM JOINED TOGETHER WILL TOTAL 8.0" INSTEAD OF 5.20" FROM END TO END. ELIMINATE THE PINCH POINT BY INCREASING THE BEND RADIUS FROM 37° TO AT LEAST 45°. MAKE YOUR DRAWING DOUBLE SIZE (2=1).
ASSEMBLY VIEW FOR REFERENCE ONLY
37° BEND
5.20
2.0" CONVEYOR CLEAT SHOWN FULL SIZE
TYPE OF OBLIQUE:
FILE: 00
GRADE:
DRAWN BY: LAST, FIRST
TEAM NO. 00
SECTION: 500
DATE DUE: 00/00/00
SCALE: FULL
IMPROVEMENT NEEDED:
☐ 1. LINE QUALITY
☐ 2. LETTERING
☐ 3. CONSTRUCTION
☐ 4. ACCURACY
32c
gev—e

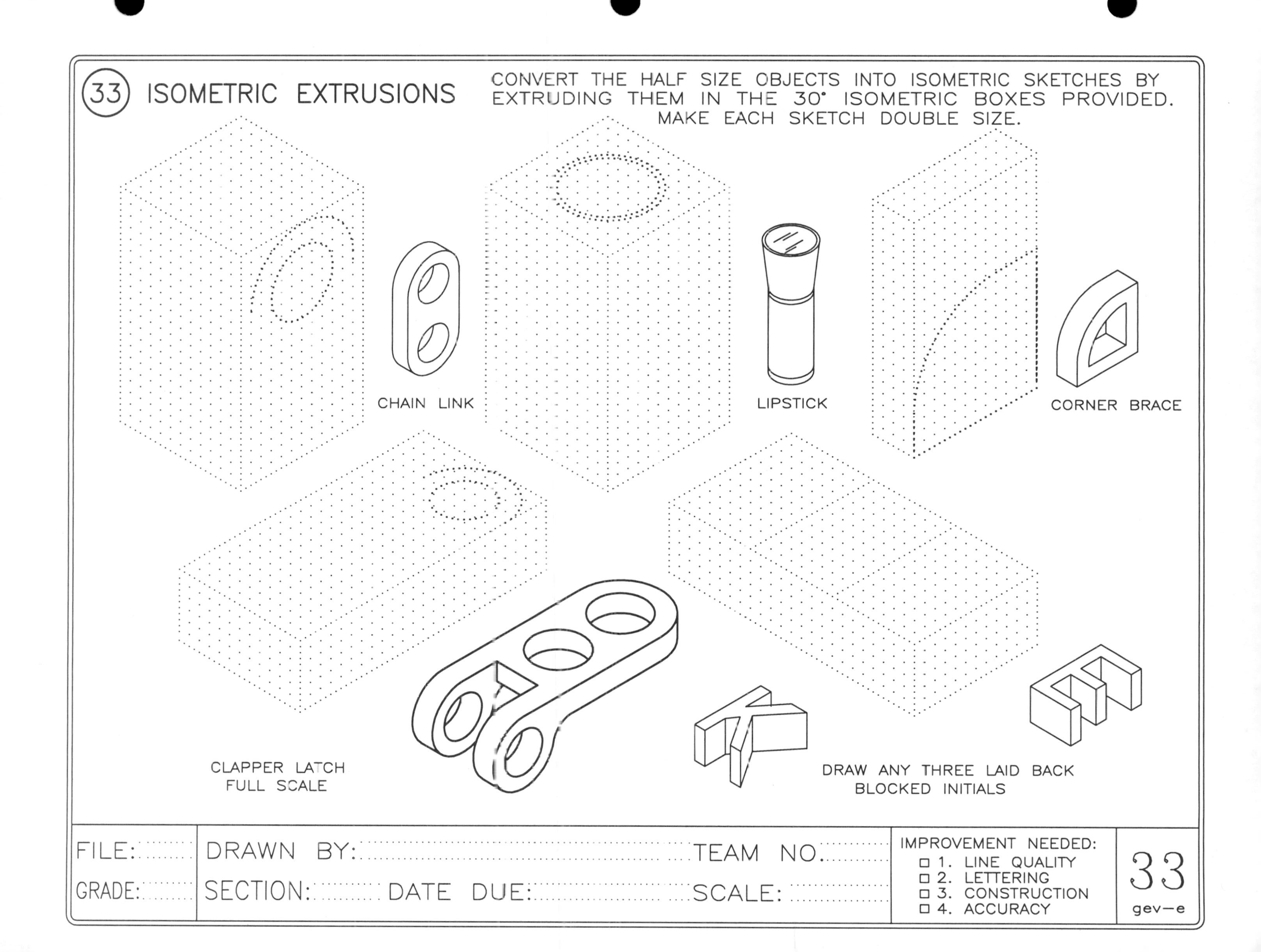

33
ISOMETRIC EXTRUSIONS
CONVERT THE HALF SIZE OBJECTS INTO ISOMETRIC SKETCHES BY EXTRUDING THEM IN THE 30° ISOMETRIC BOXES PROVIDED. MAKE EACH SKETCH DOUBLE SIZE.
CHAIN LINK
LIPSTICK
CORNER BRACE
CLAPPER LATCH
FULL SCALE
DRAW ANY THREE LAID BACK
BLOCKED INITIALS
FILE:
GRADE:
DRAWN BY:
TEAM NO.
SECTION:
DATE DUE:
SCALE:
IMPROVEMENT NEEDED:
□ 1. LINE QUALITY
□ 2. LETTERING
□ 3. CONSTRUCTION
□ 4. ACCURACY
33
gev—e

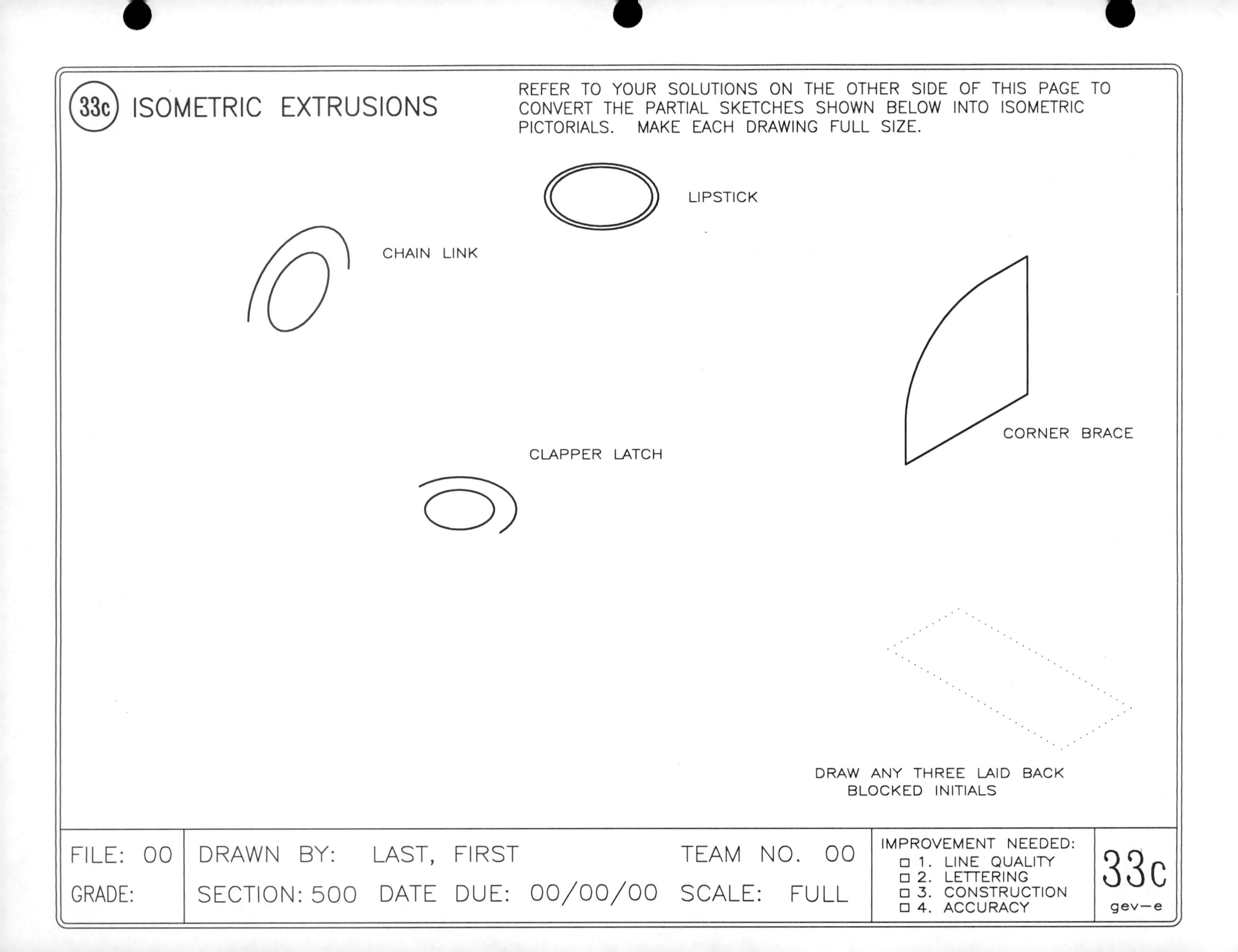
33c
ISOMETRIC EXTRUSIONS
REFER TO YOUR SOLUTIONS ON THE OTHER SIDE OF THIS PAGE TO CONVERT THE PARTIAL SKETCHES SHOWN BELOW INTO ISOMETRIC PICTORIALS. MAKE EACH DRAWING FULL SIZE.
LIPSTICK
CHAIN LINK
CORNER BRACE
CLAPPER LATCH
DRAW ANY THREE LAID BACK BLOCKED INITIALS
FILE: 00
GRADE:
DRAWN BY: LAST, FIRST
TEAM NO. 00
SECTION: 500 DATE DUE: 00/00/00 SCALE: FULL
IMPROVEMENT NEEDED:
□ 1. LINE QUALITY
□ 2. LETTERING
□ 3. CONSTRUCTION
□ 4. ACCURACY
33c
gev—e

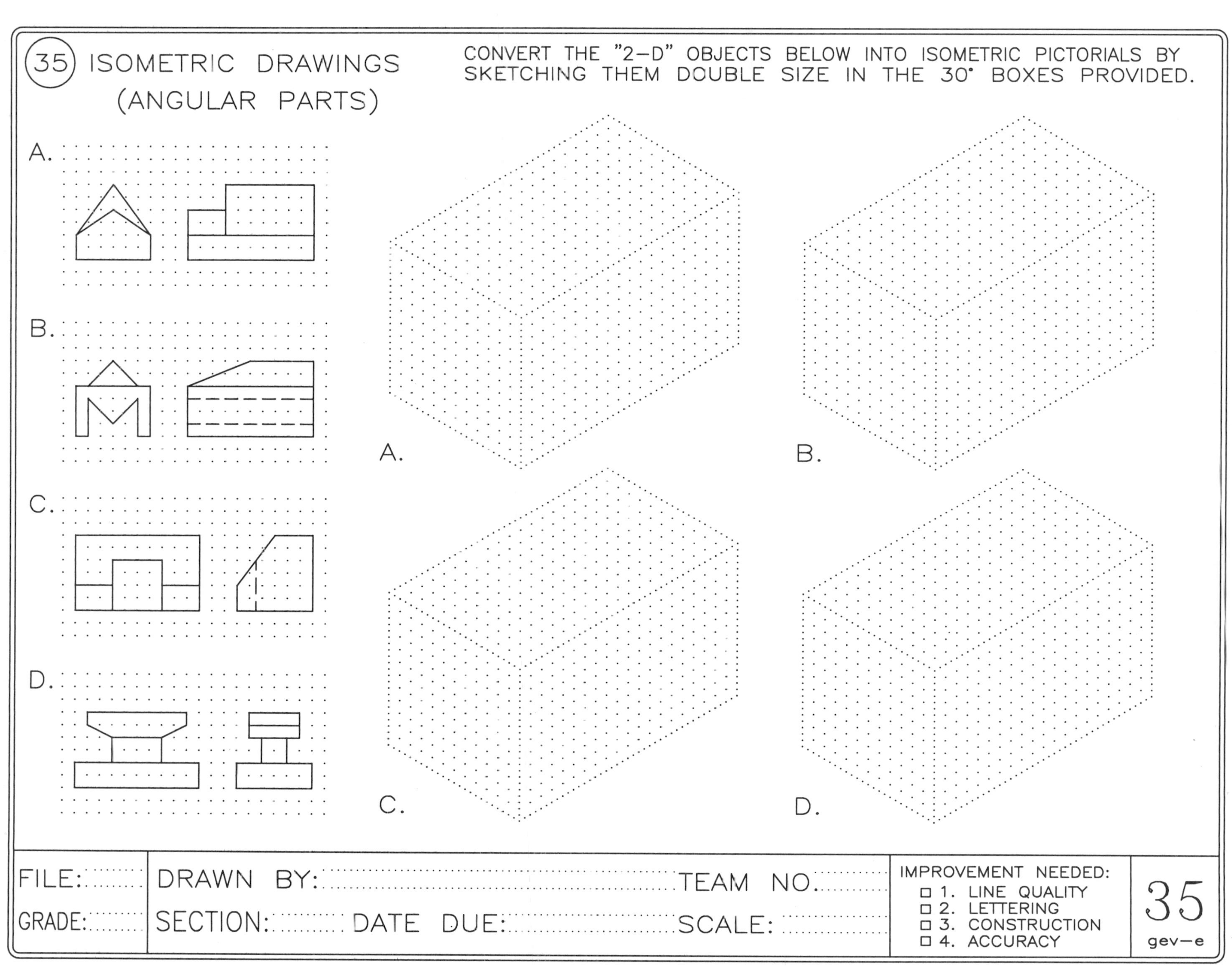
35
ISOMETRIC DRAWINGS
(ANGULAR PARTS)
CONVERT THE "2-D" OBJECTS BELOW INTO ISOMETRIC PICTORIALS BY SKETCHING THEM DOUBLE SIZE IN THE 30° BOXES PROVIDED.
A.
B.
C.
D.
A.
B.
C.
D.
FILE:
GRADE:
DRAWN BY:
TEAM NO.
SECTION:
DATE DUE:
SCALE:
IMPROVEMENT NEEDED:
☐ 1. LINE QUALITY
☐ 2. LETTERING
☐ 3. CONSTRUCTION
☐ 4. ACCURACY
35
gev-e

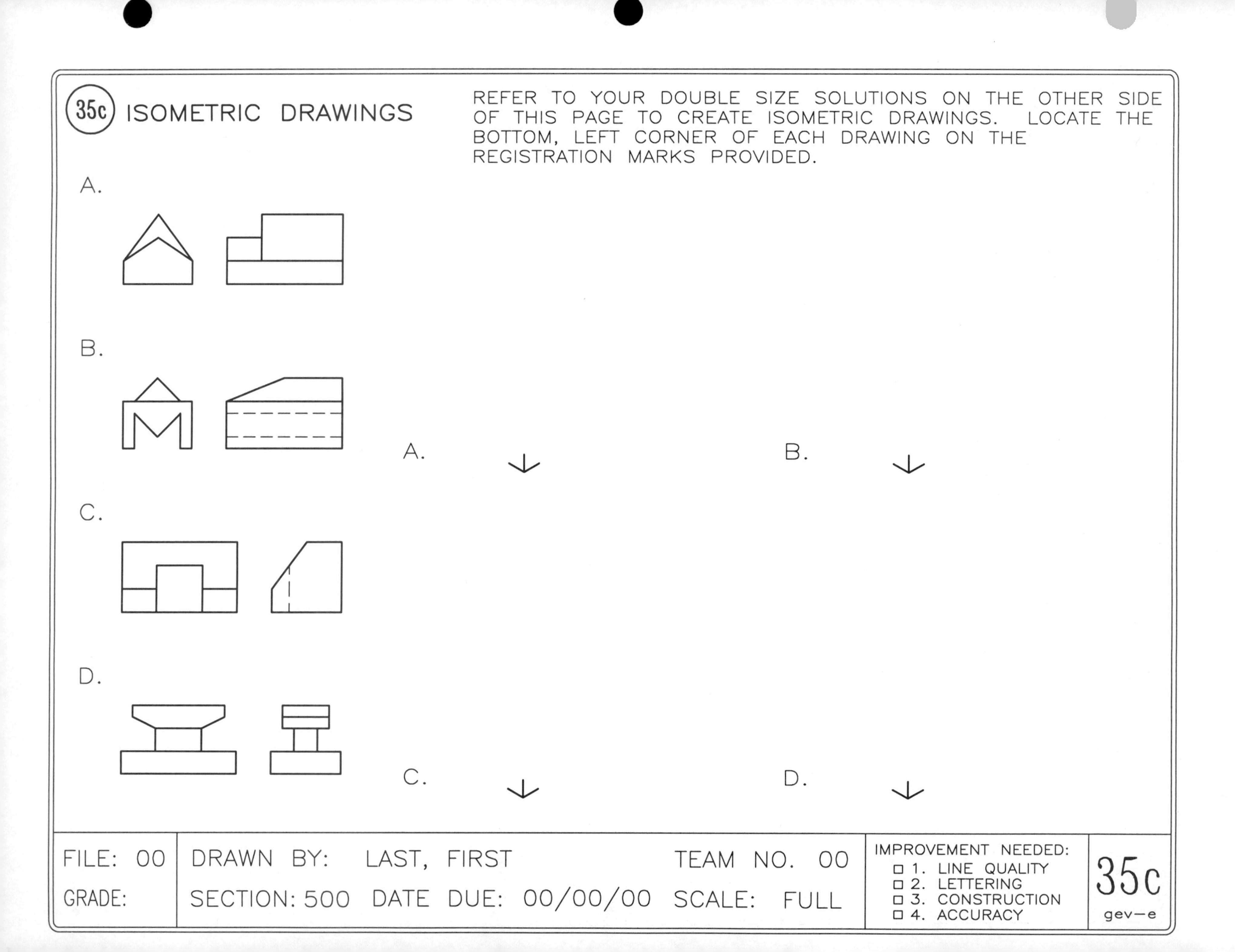
35c ISOMETRIC DRAWINGS
REFER TO YOUR DOUBLE SIZE SOLUTIONS ON THE OTHER SIDE OF THIS PAGE TO CREATE ISOMETRIC DRAWINGS. LOCATE THE BOTTOM, LEFT CORNER OF EACH DRAWING ON THE REGISTRATION MARKS PROVIDED.
A.
B.
C.
D.
A.
B.
C.
D.
FILE: 00
GRADE:
DRAWN BY: LAST, FIRST
TEAM NO. 00
SECTION: 500 DATE DUE: 00/00/00 SCALE: FULL
IMPROVEMENT NEEDED:
☐ 1. LINE QUALITY
☐ 2. LETTERING
☐ 3. CONSTRUCTION
☐ 4. ACCURACY
35c
gev—e

(36) ISOMETRIC DRAWINGS (CIRCULAR FEATURES)

CONVERT THE "2–D" OBJECTS BELOW INTO ISOMETRIC PICTORIALS BY SKETCHING THEM DOUBLE SIZE IN THE 30° BOXES PROVIDED.

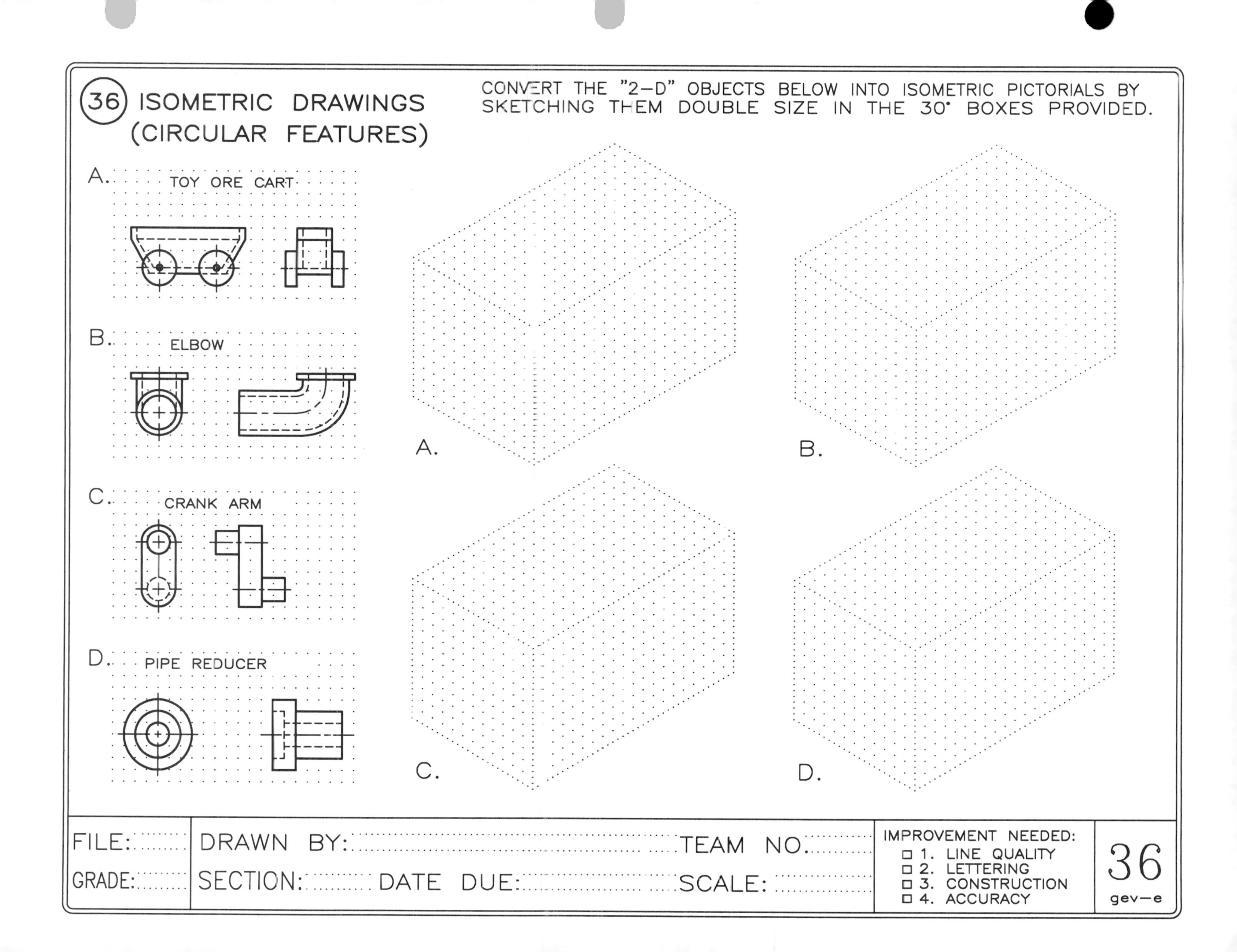

FILE:
GRADE:
DRAWN BY: TEAM NO.
SECTION: DATE DUE: SCALE:

IMPROVEMENT NEEDED:
- ☐ 1. LINE QUALITY
- ☐ 2. LETTERING
- ☐ 3. CONSTRUCTION
- ☐ 4. ACCURACY

36

gev–e

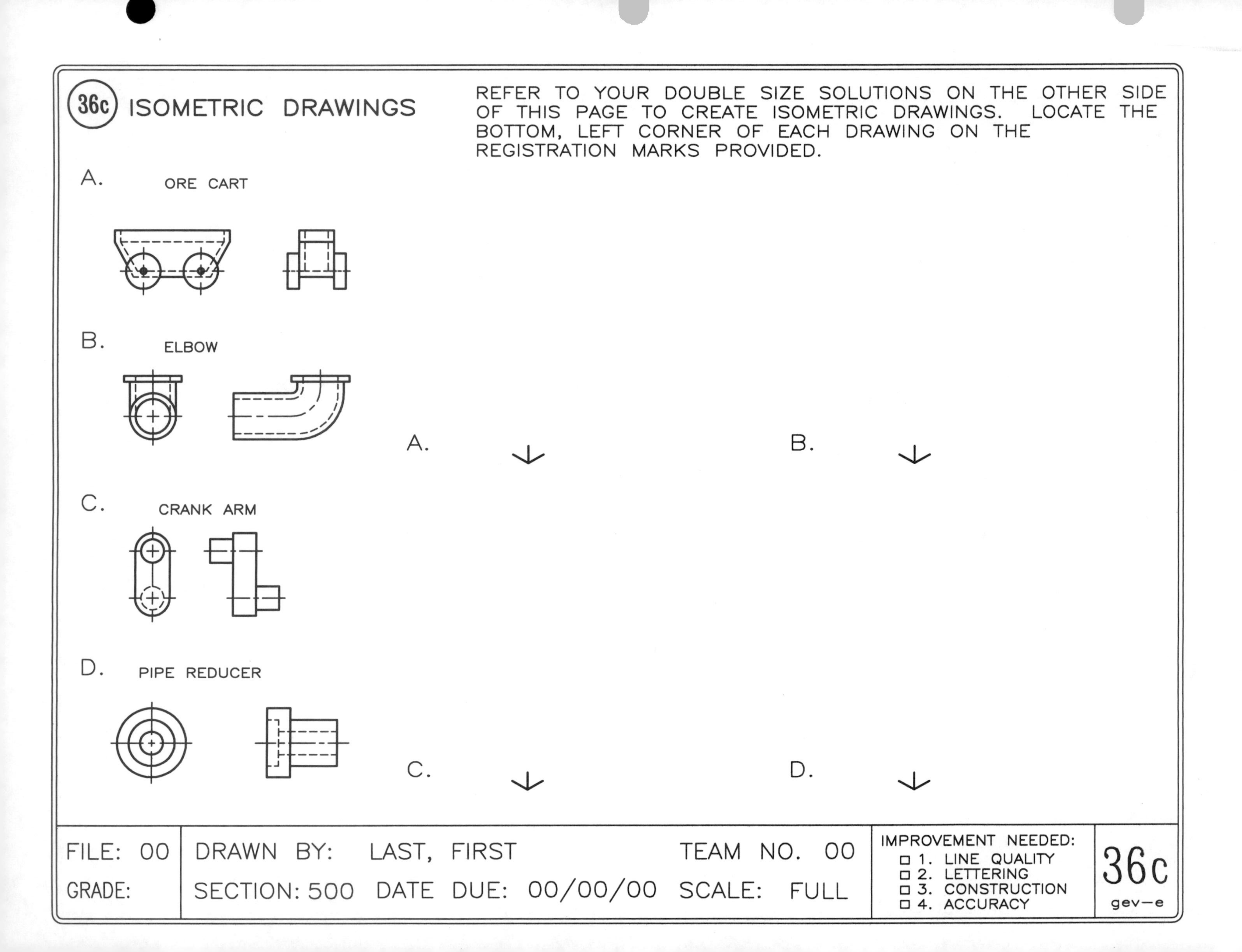
36c
ISOMETRIC DRAWINGS
REFER TO YOUR DOUBLE SIZE SOLUTIONS ON THE OTHER SIDE OF THIS PAGE TO CREATE ISOMETRIC DRAWINGS. LOCATE THE BOTTOM, LEFT CORNER OF EACH DRAWING ON THE REGISTRATION MARKS PROVIDED.
A. ORE CART
B. ELBOW
C. CRANK ARM
D. PIPE REDUCER
A.
B.
C.
D.
FILE: 00
GRADE:
DRAWN BY: LAST, FIRST
TEAM NO. 00
SECTION: 500
DATE DUE: 00/00/00
SCALE: FULL
IMPROVEMENT NEEDED:
1. LINE QUALITY
2. LETTERING
3. CONSTRUCTION
4. ACCURACY
36c
gev-e

(37) ISOMETRIC DRAWINGS

CONVERT THE HALF SIZE, TWO VIEW DRAWING IN FIGURE "H" INTO A HALF SIZE ISOMETRIC PICTORIAL SKETCH.

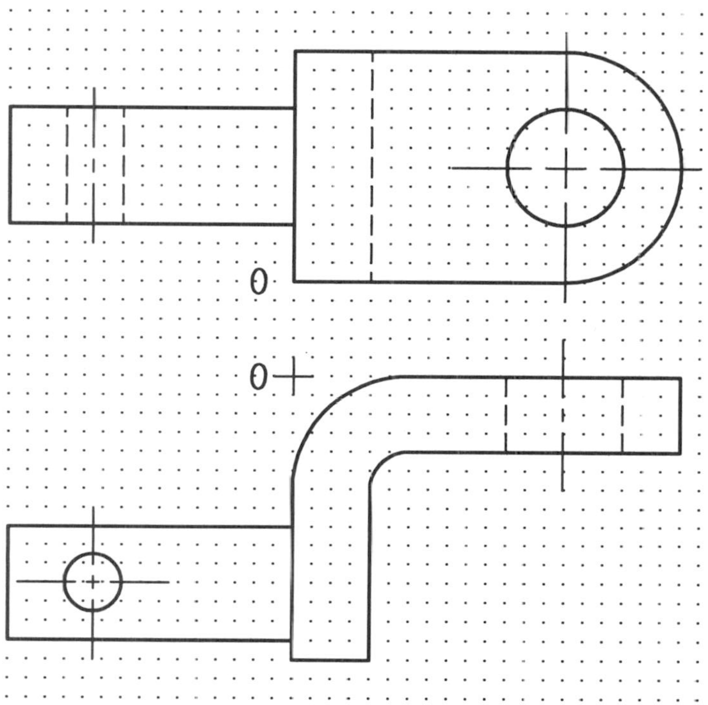

FIGURE H: "HITCH TONGUE"
HALF SCALE (1=2)

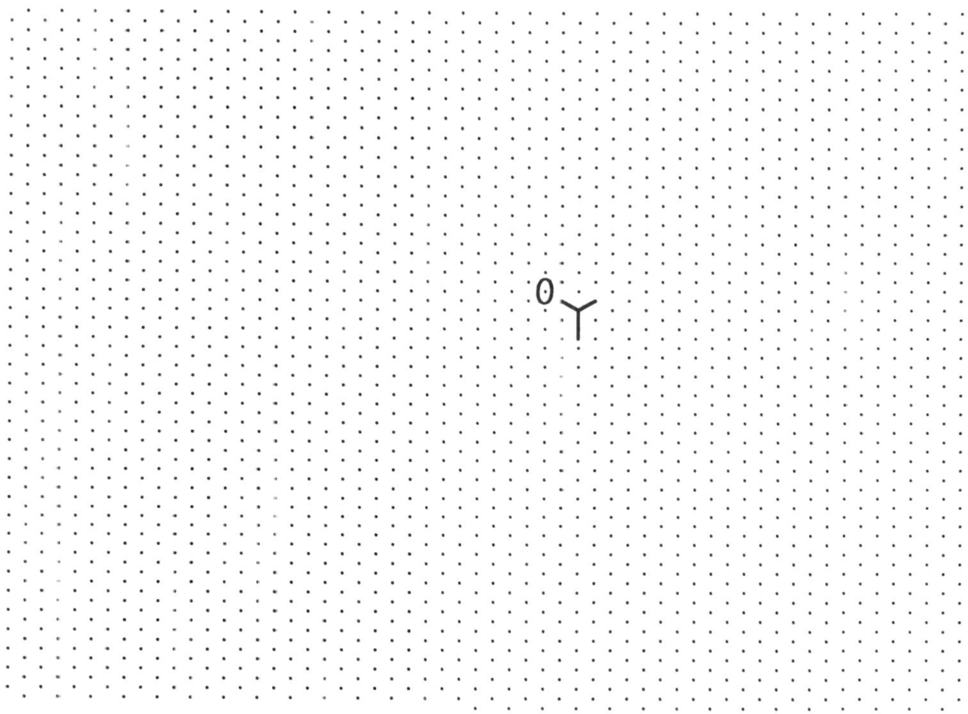

FILE:	DRAWN BY:	TEAM NO.	IMPROVEMENT NEEDED: ☐ 1. LINE QUALITY ☐ 2. LETTERING ☐ 3. CONSTRUCTION ☐ 4. ACCURACY	37
GRADE:	SECTION: DATE DUE:	SCALE:		gev-e

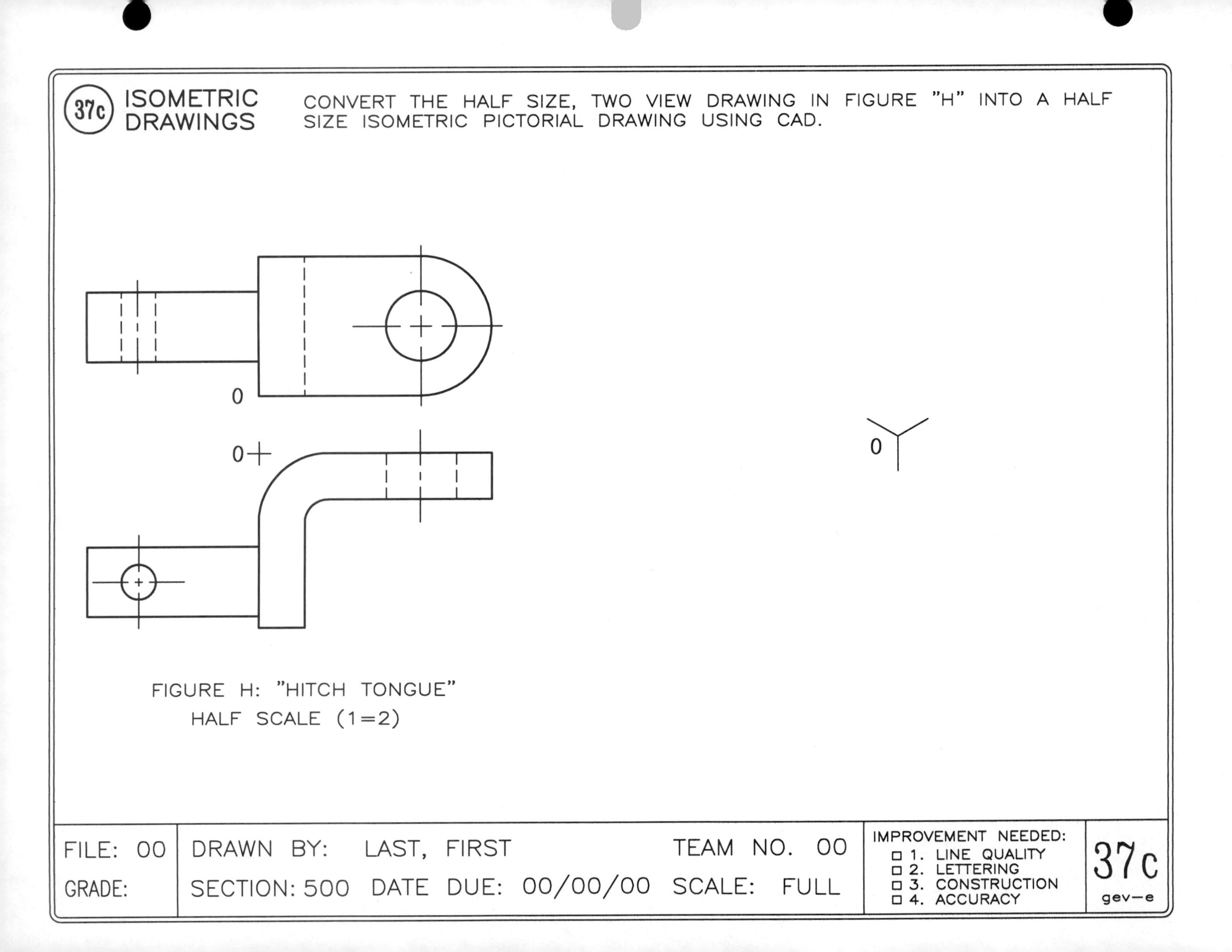
37c
ISOMETRIC DRAWINGS
CONVERT THE HALF SIZE, TWO VIEW DRAWING IN FIGURE "H" INTO A HALF SIZE ISOMETRIC PICTORIAL DRAWING USING CAD.
0
0
0
FIGURE H: "HITCH TONGUE"
HALF SCALE (1=2)
FILE: 00
GRADE:
DRAWN BY: LAST, FIRST
TEAM NO. 00
SECTION: 500
DATE DUE: 00/00/00
SCALE: FULL
IMPROVEMENT NEEDED:
☐ 1. LINE QUALITY
☐ 2. LETTERING
☐ 3. CONSTRUCTION
☐ 4. ACCURACY
37c
gev-e

(38) ISOMETRIC DRAWINGS

CONVERT THE TWO VIEW DRAWING OF THE MAGNESIUM ROBOT FOOT SHOWN IN FIGURE "J" INTO A FULL SIZE ISOMETRIC PICTORIAL SKETCH.

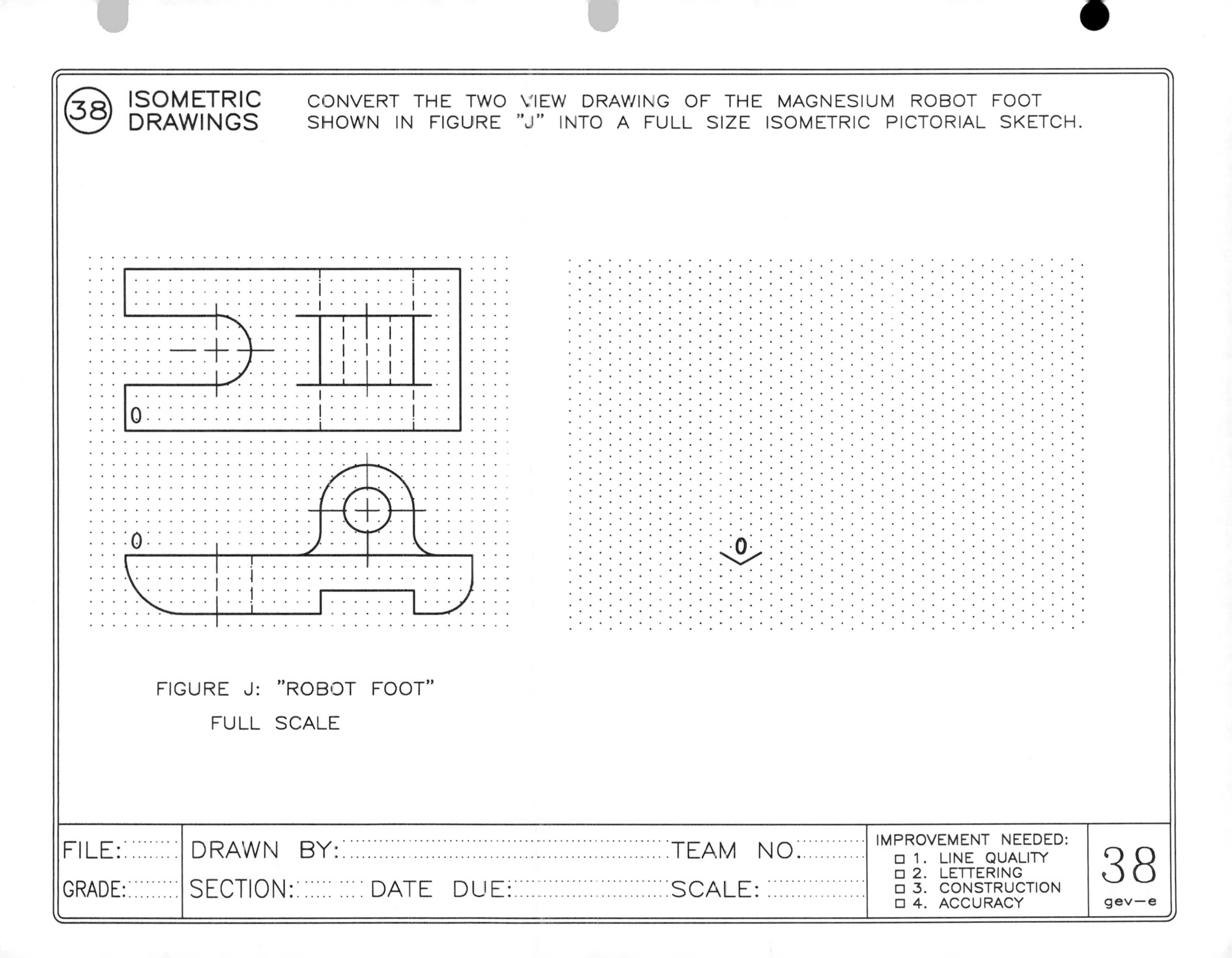

FIGURE J: "ROBOT FOOT"

FULL SCALE

FILE:	DRAWN BY:	TEAM NO.	IMPROVEMENT NEEDED: ☐ 1. LINE QUALITY ☐ 2. LETTERING ☐ 3. CONSTRUCTION ☐ 4. ACCURACY	38
GRADE:	SECTION: DATE DUE:	SCALE:		gev–e

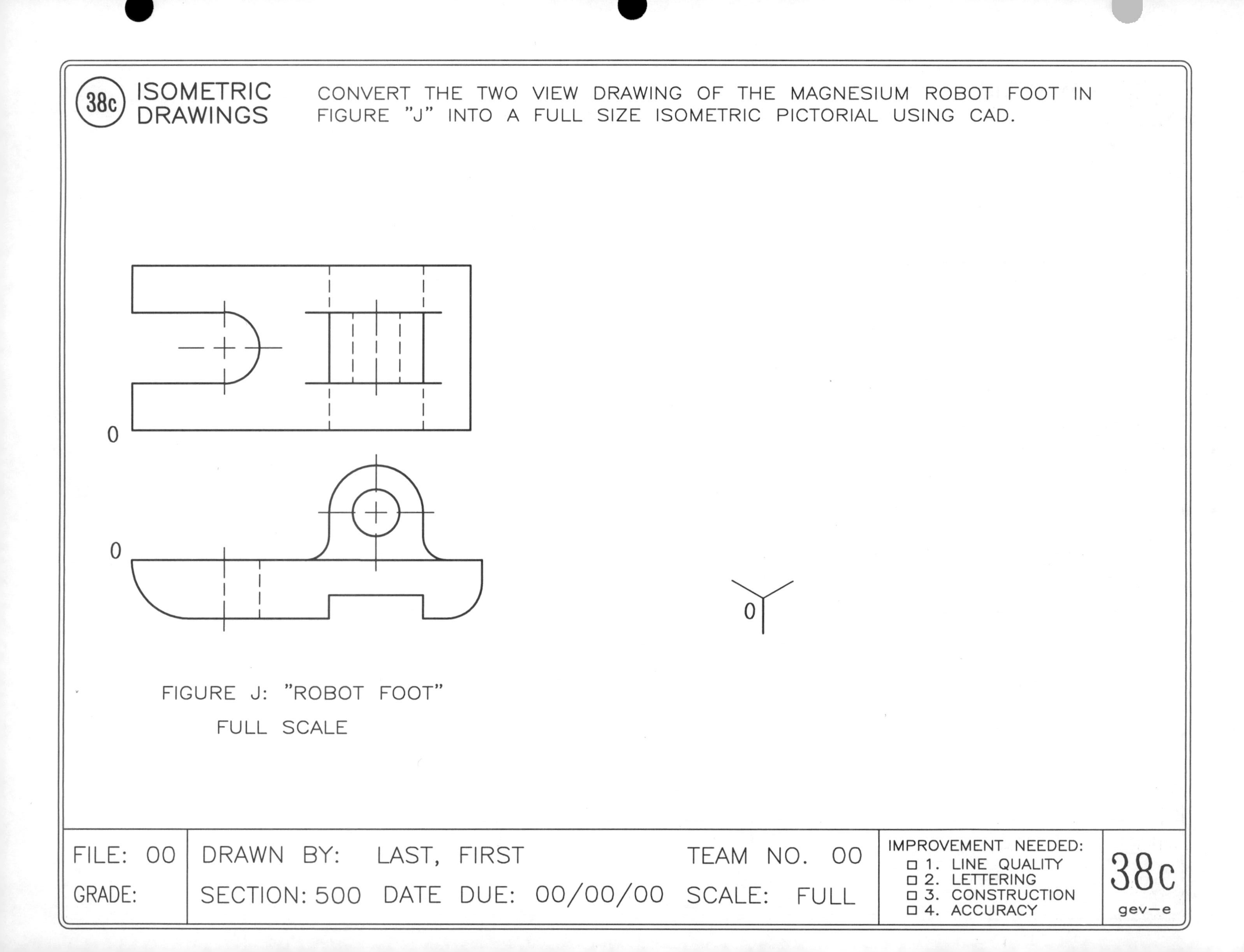

38c
ISOMETRIC DRAWINGS
CONVERT THE TWO VIEW DRAWING OF THE MAGNESIUM ROBOT FOOT IN FIGURE "J" INTO A FULL SIZE ISOMETRIC PICTORIAL USING CAD.
0
0
0
FIGURE J: "ROBOT FOOT"
FULL SCALE
FILE: 00
GRADE:
DRAWN BY: LAST, FIRST
TEAM NO. 00
SECTION: 500 DATE DUE: 00/00/00 SCALE: FULL
IMPROVEMENT NEEDED:
1. LINE QUALITY
2. LETTERING
3. CONSTRUCTION
4. ACCURACY
38c
gev–e

(39) LETTERING MASTERY

COPY EACH SENTENCE ONCE OR TWICE AS ASSIGNED BY YOUR INSTRUCTOR. FRACTICE PROPER SPACING BETWEEN LETTERS, LINES AND WORDS. USE A SOFT, MEDIUM POINTED LEAD, AND ALLOW A FULL SPACE BETWEEN EACH SENTENCE.

1. CROSSHATCH PATTERNS ARE DRAWN WITH A THIN (.010") LINETYPE.
2. TYPES OF METAL ARE INDICATED BY STANDARDIZED "ANSI" PATTERNS.
3. AVOID SETTING HATCH ANGLES THAT APPEAR VERTICAL OR HORIZONTAL.
4. HALF SECTIONS ACTUALLY REMOVE ONLY ONE FOURTH OF THE OBJECT.
5. BROKEN—OUT SECTIONS CAN REVEAL NEAR OR DISTANT FEATURES.
6. REVOLVED SECTIONS REVEAL BOTH INTERIOR AND EXTERIOR DETAILS.
7. CUTTING PLANES ARE MADE WITH A THICK (.020") PHANTOM LINETYPE.

FILE:	DRAWN BY:	TEAM NO.	IMPROVEMENT NEEDED: ☐ 1. LINE QUALITY ☐ 2. LETTERING ☐ 3. CONSTRUCTION ☐ 4. ACCURACY	39 gev—e
GRADE:	SECTION: DATE DUE:	SCALE:		

39c

LETTERING MASTERY

TYPE EACH MEMORY LINE FROM THE OTHER SIDE OF THIS PAGE ONCE OR TWICE AS ASSIGNED BY YOUR INSTRUCTOR. USE A DIFFERENT TEXT STYLE AND FONT FOR EACH OF THE SEVEN SENTENCES.

1.

2.

3.

4.

5.

6.

7.

FILE: 00	DRAWN BY: LAST, FIRST	TEAM NO. 00	IMPROVEMENT NEEDED:	39c
GRADE:	SECTION: 500 DATE DUE: 00/00/00	SCALE: FULL	☐ 1. LINE QUALITY ☐ 2. LETTERING ☐ 3. CONSTRUCTION ☐ 4. ACCURACY	gev–e

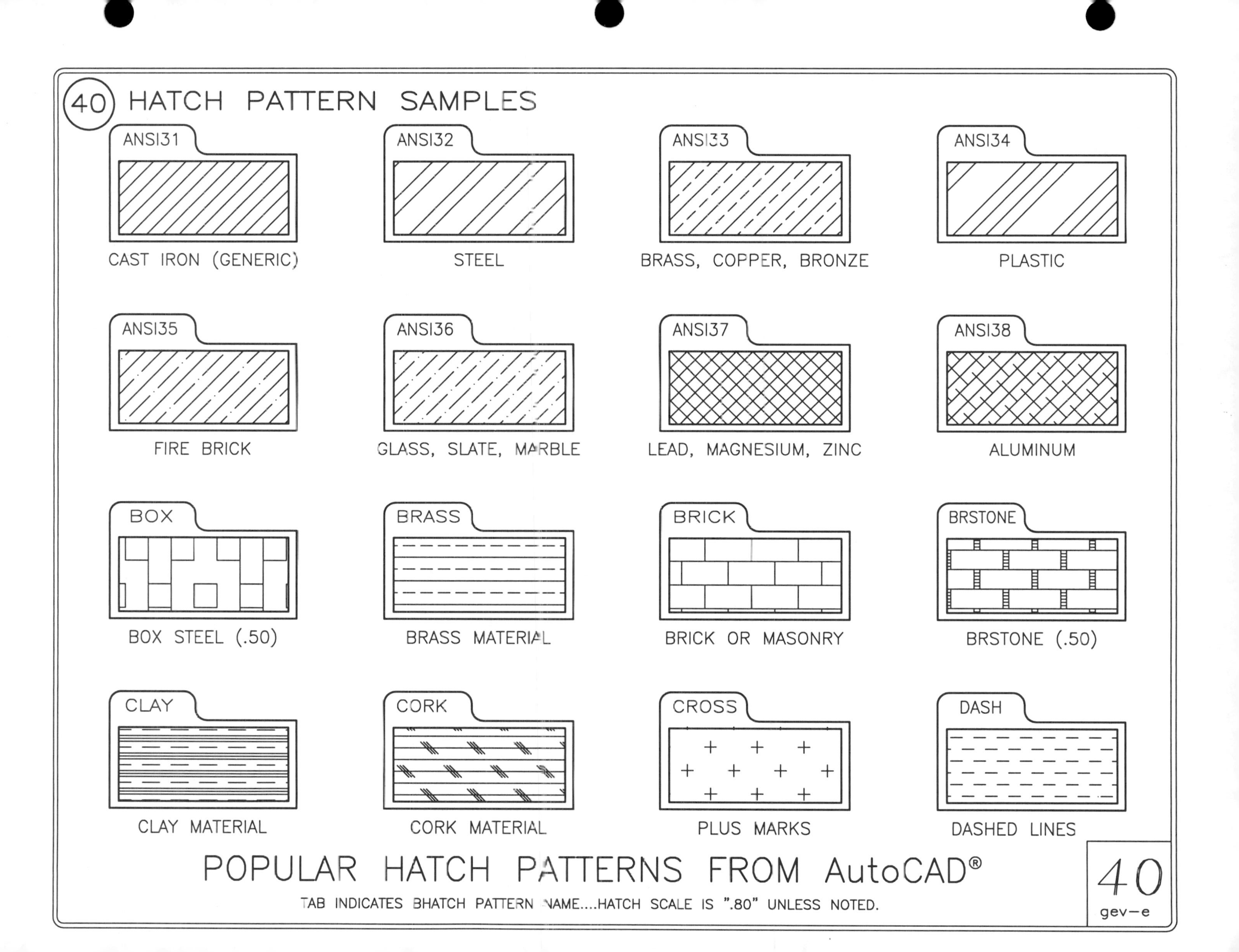
40
HATCH PATTERN SAMPLES
ANSI31
CAST IRON (GENERIC)
ANSI32
STEEL
ANSI33
BRASS, COPPER, BRONZE
ANSI34
PLASTIC
ANSI35
FIRE BRICK
ANSI36
GLASS, SLATE, MARBLE
ANSI37
LEAD, MAGNESIUM, ZINC
ANSI38
ALUMINUM
BOX
BOX STEEL (.50)
BRASS
BRASS MATERIAL
BRICK
BRICK OR MASONRY
BRSTONE
BRSTONE (.50)
CLAY
CLAY MATERIAL
CORK
CORK MATERIAL
CROSS
PLUS MARKS
DASH
DASHED LINES
POPULAR HATCH PATTERNS FROM AutoCAD®
TAB INDICATES BHATCH PATTERN NAME....HATCH SCALE IS ".80" UNLESS NOTED.
40
gev-e

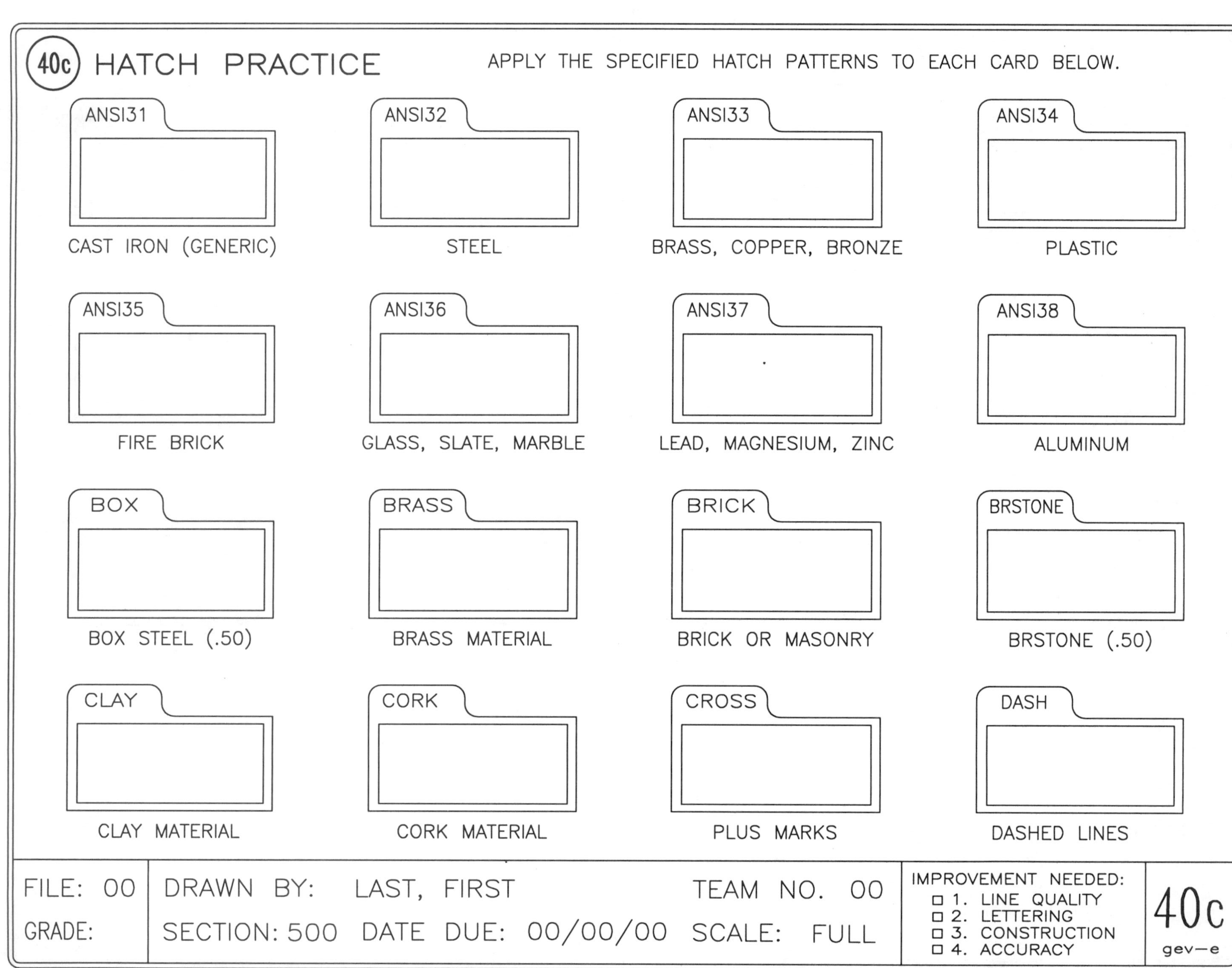
40c
HATCH PRACTICE
APPLY THE SPECIFIED HATCH PATTERNS TO EACH CARD BELOW.
ANSI31
CAST IRON (GENERIC)
ANSI32
STEEL
ANSI33
BRASS, COPPER, BRONZE
ANSI34
PLASTIC
ANSI35
FIRE BRICK
ANSI36
GLASS, SLATE, MARBLE
ANSI37
LEAD, MAGNESIUM, ZINC
ANSI38
ALUMINUM
BOX
BOX STEEL (.50)
BRASS
BRASS MATERIAL
BRICK
BRICK OR MASONRY
BRSTONE
BRSTONE (.50)
CLAY
CLAY MATERIAL
CORK
CORK MATERIAL
CROSS
PLUS MARKS
DASH
DASHED LINES
FILE: 00
GRADE:
DRAWN BY: LAST, FIRST
TEAM NO. 00
SECTION: 500 DATE DUE: 00/00/00 SCALE: FULL
IMPROVEMENT NEEDED:
□ 1. LINE QUALITY
□ 2. LETTERING
□ 3. CONSTRUCTION
□ 4. ACCURACY
40c
gev—e

41 HATCH PATTERN SAMPLES

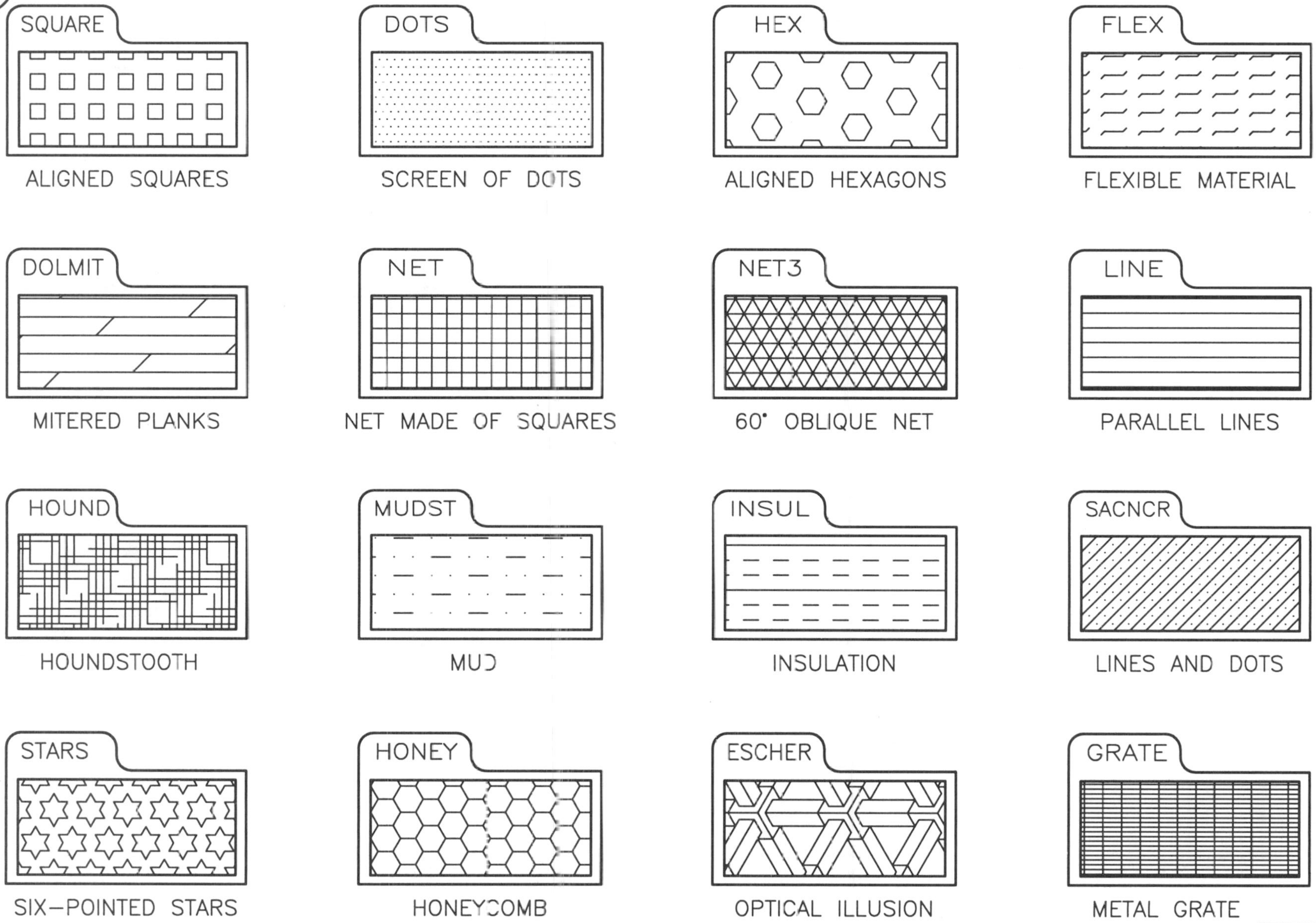

Tab	Pattern	Tab	Pattern	Tab	Pattern	Tab	Pattern
SQUARE	ALIGNED SQUARES	DOTS	SCREEN OF DOTS	HEX	ALIGNED HEXAGONS	FLEX	FLEXIBLE MATERIAL
DOLMIT	MITERED PLANKS	NET	NET MADE OF SQUARES	NET3	60° OBLIQUE NET	LINE	PARALLEL LINES
HOUND	HOUNDSTOOTH	MUDST	MUD	INSUL	INSULATION	SACNCR	LINES AND DOTS
STARS	SIX-POINTED STARS	HONEY	HONEYCOMB	ESCHER	OPTICAL ILLUSION	GRATE	METAL GRATE

POPULAR HATCH PATTERNS FROM AutoCAD®

TAB INDICATES BHATCH PATTERN NAME....HATCH SCALE VARIES FROM .50 TO 1.0 TO FIT.

41
gev-e

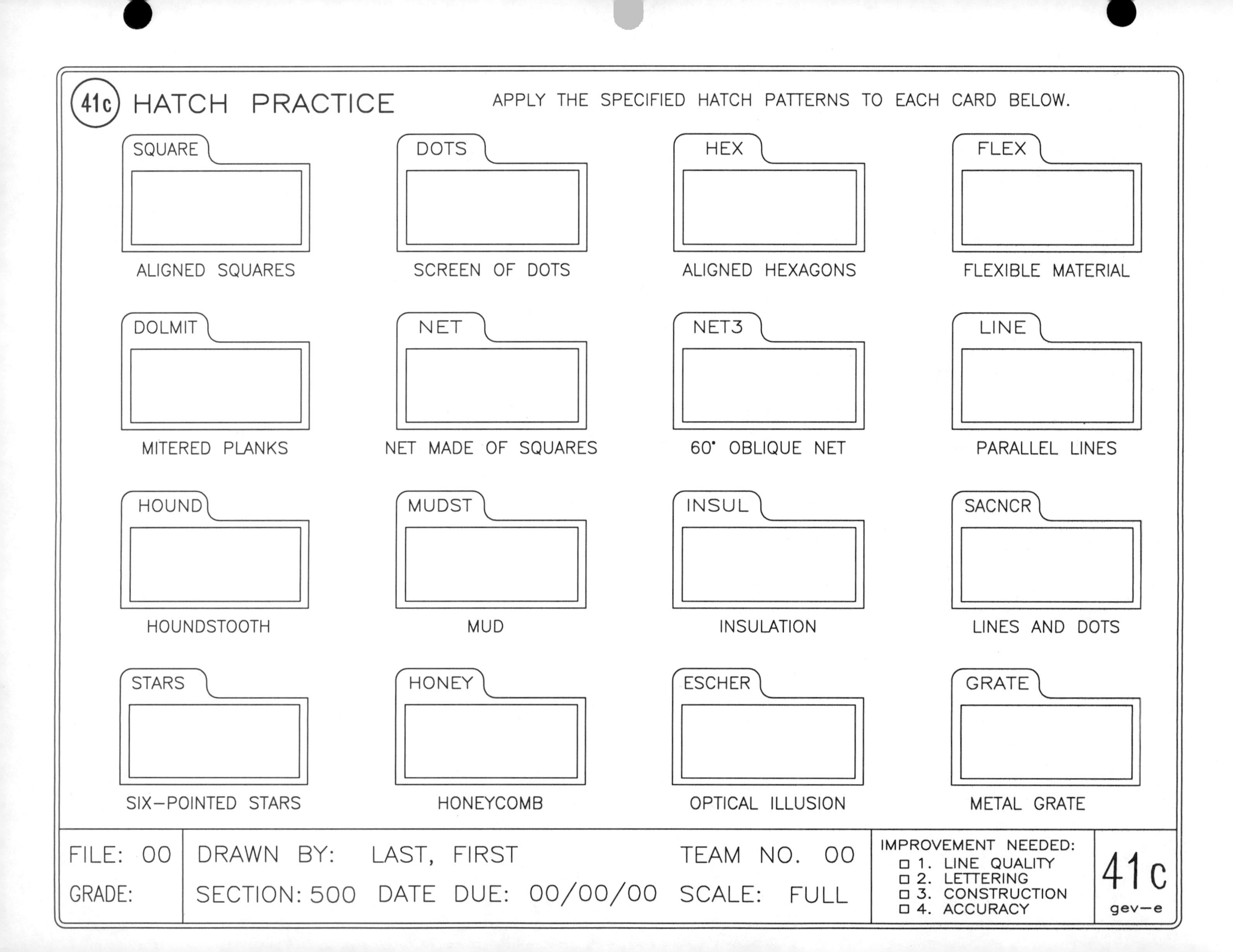
41c
HATCH PRACTICE
APPLY THE SPECIFIED HATCH PATTERNS TO EACH CARD BELOW.
SQUARE
ALIGNED SQUARES
DOTS
SCREEN OF DOTS
HEX
ALIGNED HEXAGONS
FLEX
FLEXIBLE MATERIAL
DOLMIT
MITERED PLANKS
NET
NET MADE OF SQUARES
NET3
60° OBLIQUE NET
LINE
PARALLEL LINES
HOUND
HOUNDSTOOTH
MUDST
MUD
INSUL
INSULATION
SACNCR
LINES AND DOTS
STARS
SIX-POINTED STARS
HONEY
HONEYCOMB
ESCHER
OPTICAL ILLUSION
GRATE
METAL GRATE
FILE: 00
GRADE:
DRAWN BY: LAST, FIRST
TEAM NO. 00
SECTION: 500
DATE DUE: 00/00/00
SCALE: FULL
IMPROVEMENT NEEDED:
☐ 1. LINE QUALITY
☐ 2. LETTERING
☐ 3. CONSTRUCTION
☐ 4. ACCURACY
41c
gev-e

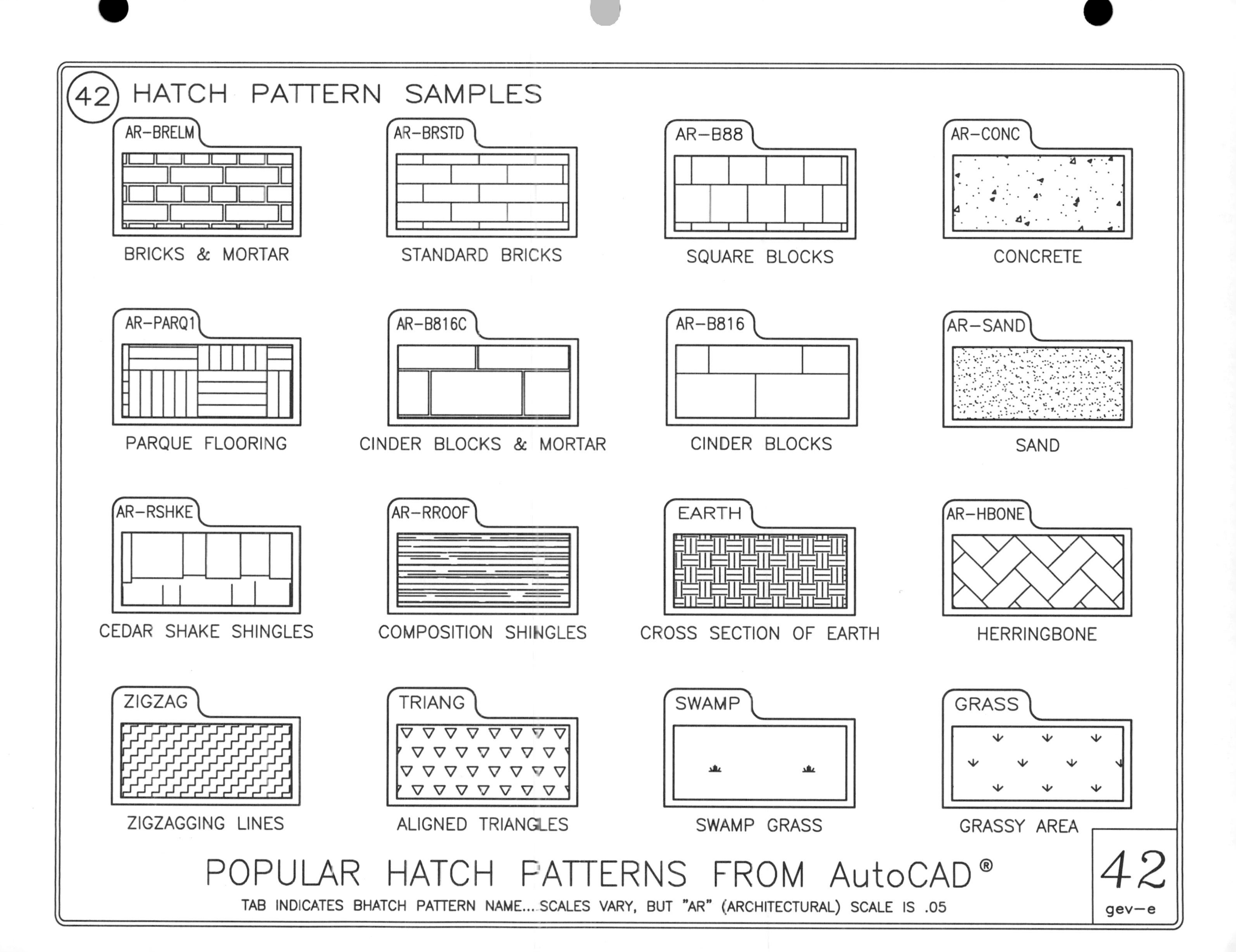
42
HATCH PATTERN SAMPLES
AR–BRELM
BRICKS & MORTAR
AR–BRSTD
STANDARD BRICKS
AR–B88
SQUARE BLOCKS
AR–CONC
CONCRETE
AR–PARQ1
PARQUE FLOORING
AR–B816C
CINDER BLOCKS & MORTAR
AR–B816
CINDER BLOCKS
AR–SAND
SAND
AR–RSHKE
CEDAR SHAKE SHINGLES
AR–RROOF
COMPOSITION SHINGLES
EARTH
CROSS SECTION OF EARTH
AR–HBONE
HERRINGBONE
ZIGZAG
ZIGZAGGING LINES
TRIANG
ALIGNED TRIANGLES
SWAMP
SWAMP GRASS
GRASS
GRASSY AREA
POPULAR HATCH PATTERNS FROM AutoCAD®
TAB INDICATES BHATCH PATTERN NAME....SCALES VARY, BUT "AR" (ARCHITECTURAL) SCALE IS .05
42
gev–e

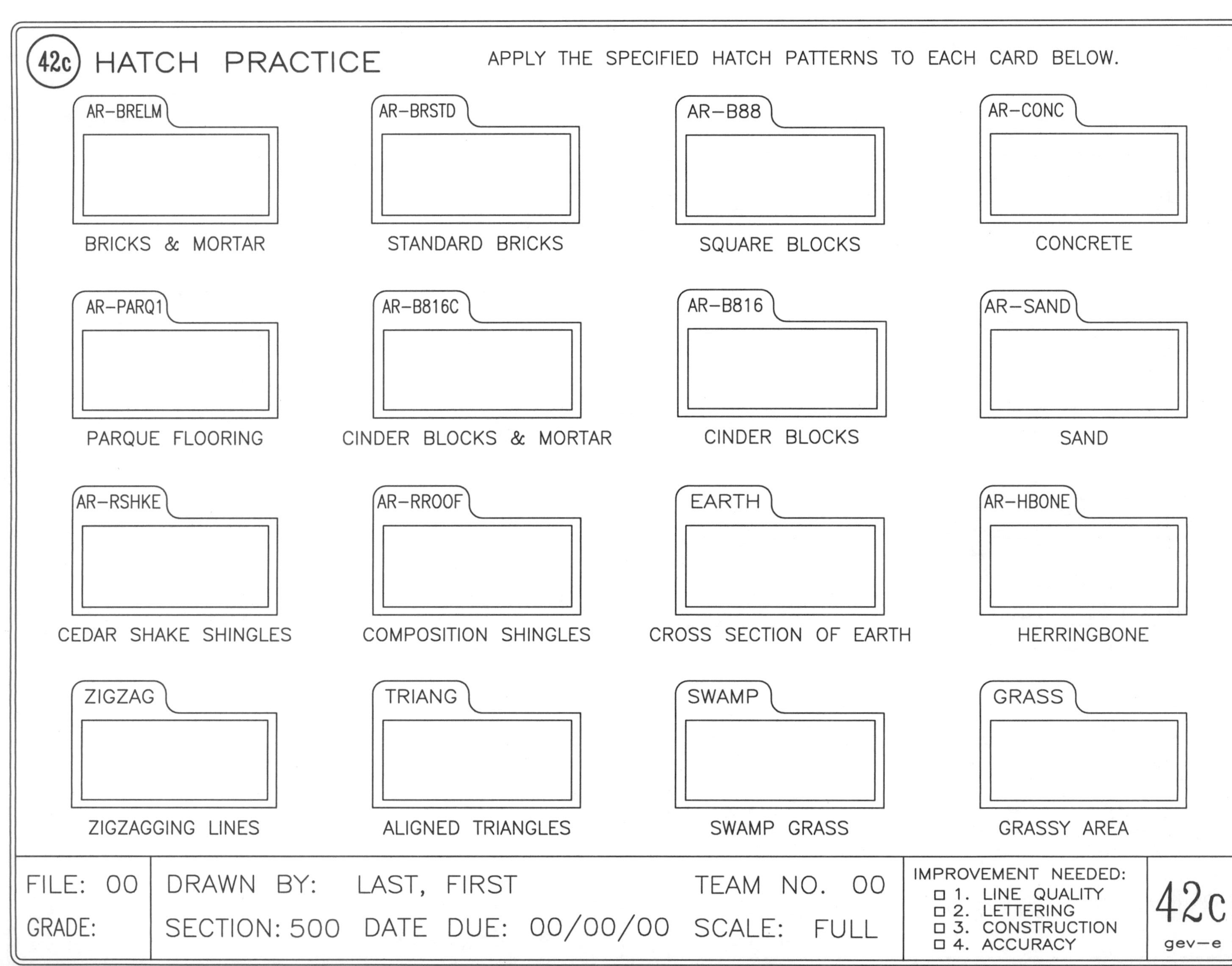

42c HATCH PRACTICE
APPLY THE SPECIFIED HATCH PATTERNS TO EACH CARD BELOW.
AR-BRELM
BRICKS & MORTAR
AR-BRSTD
STANDARD BRICKS
AR-B88
SQUARE BLOCKS
AR-CONC
CONCRETE
AR-PARQ1
PARQUE FLOORING
AR-B816C
CINDER BLOCKS & MORTAR
AR-B816
CINDER BLOCKS
AR-SAND
SAND
AR-RSHKE
CEDAR SHAKE SHINGLES
AR-RROOF
COMPOSITION SHINGLES
EARTH
CROSS SECTION OF EARTH
AR-HBONE
HERRINGBONE
ZIGZAG
ZIGZAGGING LINES
TRIANG
ALIGNED TRIANGLES
SWAMP
SWAMP GRASS
GRASS
GRASSY AREA
FILE: 00
GRADE:
DRAWN BY: LAST, FIRST
TEAM NO. 00
SECTION: 500
DATE DUE: 00/00/00
SCALE: FULL
IMPROVEMENT NEEDED:
☐ 1. LINE QUALITY
☐ 2. LETTERING
☐ 3. CONSTRUCTION
☐ 4. ACCURACY
42c
gev-e

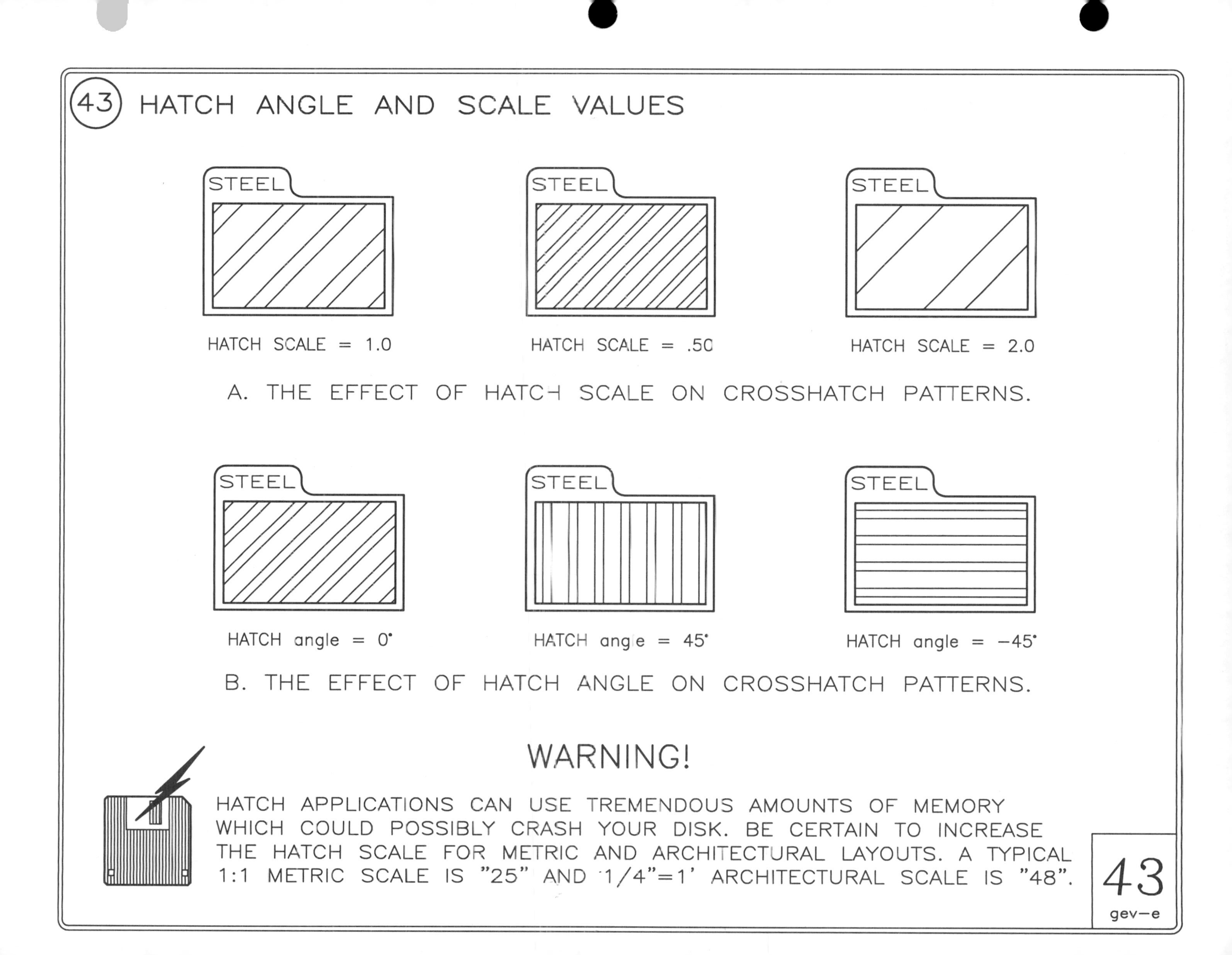
43
HATCH ANGLE AND SCALE VALUES
STEEL
HATCH SCALE = 1.0
STEEL
HATCH SCALE = .50
STEEL
HATCH SCALE = 2.0
A. THE EFFECT OF HATCH SCALE ON CROSSHATCH PATTERNS.
STEEL
HATCH angle = 0°
STEEL
HATCH angle = 45°
STEEL
HATCH angle = −45°
B. THE EFFECT OF HATCH ANGLE ON CROSSHATCH PATTERNS.
WARNING!
HATCH APPLICATIONS CAN USE TREMENDOUS AMOUNTS OF MEMORY
WHICH COULD POSSIBLY CRASH YOUR DISK. BE CERTAIN TO INCREASE
THE HATCH SCALE FOR METRIC AND ARCHITECTURAL LAYOUTS. A TYPICAL
1:1 METRIC SCALE IS "25" AND 1/4"=1' ARCHITECTURAL SCALE IS "48".
43
gev−e

43c HATCH PRACTICE

APPLY THE SPECIFIED HATCH PATTERNS TO EACH CARD BELOW. CUSTOMIZE THE HATCH SCALE AND ANGLE AS SPECIFIED.

STEEL — HATCH SCALE =1.0

STEEL — HATCH SCALE =.50

STEEL — HATCH SCALE =2.0

A. THE EFFECT OF HATCH SCALE ON CROSSHATCH PATTERNS.

STEEL — HATCH angle = 0°

STEEL — HATCH angle = 45°

STEEL — HATCH angle = −45°

B. THE EFFECT OF HATCH ANGLE ON CROSSHATCH PATTERNS.

WARNING!

HATCH APPLICATIONS CAN USE TREMENDOUS AMOUNTS OF MEMORY WHICH COULD POSSIBLY CRASH YOUR DISK. BE CERTAIN TO INCREASE THE HATCH SCALE FOR METRIC AND ARCHITECTURAL LAYOUTS. A TYPICAL 1:1 METRIC SCALE IS "25" AND 1/4"=1' ARCHITECTURAL SCALE IS "48".

FILE: 00	DRAWN BY: LAST, FIRST	TEAM NO. 00	IMPROVEMENT NEEDED:	43c
GRADE:	SECTION: 500 DATE DUE: 00/00/00	SCALE: FULL	☐ 1. LINE QUALITY ☐ 2. LETTERING ☐ 3. CONSTRUCTION ☐ 4. ACCURACY	gev–e

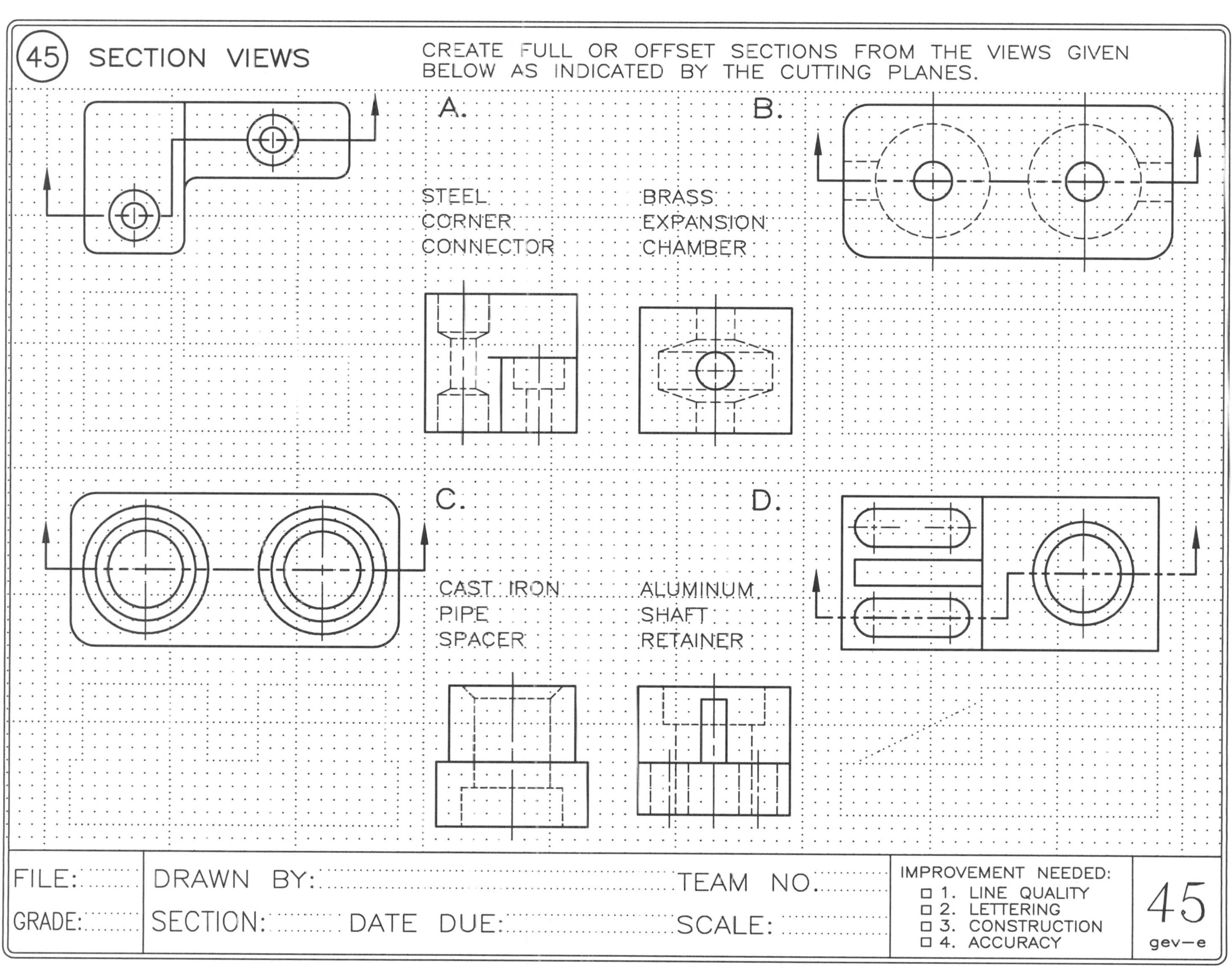

45
SECTION VIEWS
CREATE FULL OR OFFSET SECTIONS FROM THE VIEWS GIVEN BELOW AS INDICATED BY THE CUTTING PLANES.
A.
STEEL
CORNER
CONNECTOR
B.
BRASS
EXPANSION
CHAMBER
C.
CAST IRON
PIPE
SPACER
D.
ALUMINUM
SHAFT
RETAINER
FILE:
GRADE:
DRAWN BY:
SECTION:
DATE DUE:
TEAM NO.
SCALE:
IMPROVEMENT NEEDED:
☐ 1. LINE QUALITY
☐ 2. LETTERING
☐ 3. CONSTRUCTION
☐ 4. ACCURACY
45
gev—e

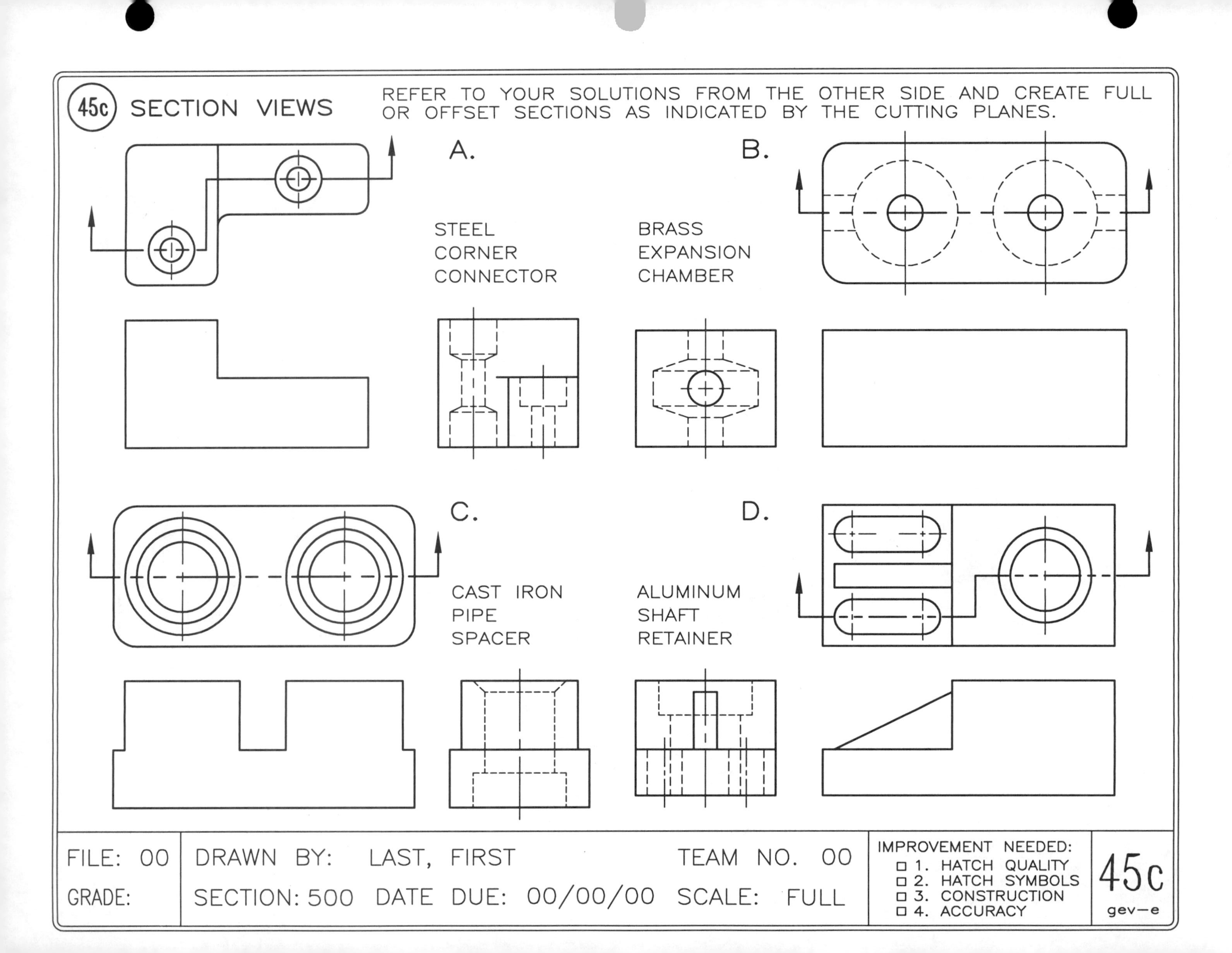
45c
SECTION VIEWS
REFER TO YOUR SOLUTIONS FROM THE OTHER SIDE AND CREATE FULL OR OFFSET SECTIONS AS INDICATED BY THE CUTTING PLANES.
A.
STEEL CORNER CONNECTOR
B.
BRASS EXPANSION CHAMBER
C.
CAST IRON PIPE SPACER
D.
ALUMINUM SHAFT RETAINER
FILE: 00
GRADE:
DRAWN BY: LAST, FIRST
TEAM NO. 00
SECTION: 500
DATE DUE: 00/00/00
SCALE: FULL
IMPROVEMENT NEEDED:
□ 1. HATCH QUALITY
□ 2. HATCH SYMBOLS
□ 3. CONSTRUCTION
□ 4. ACCURACY
45c
gev—e

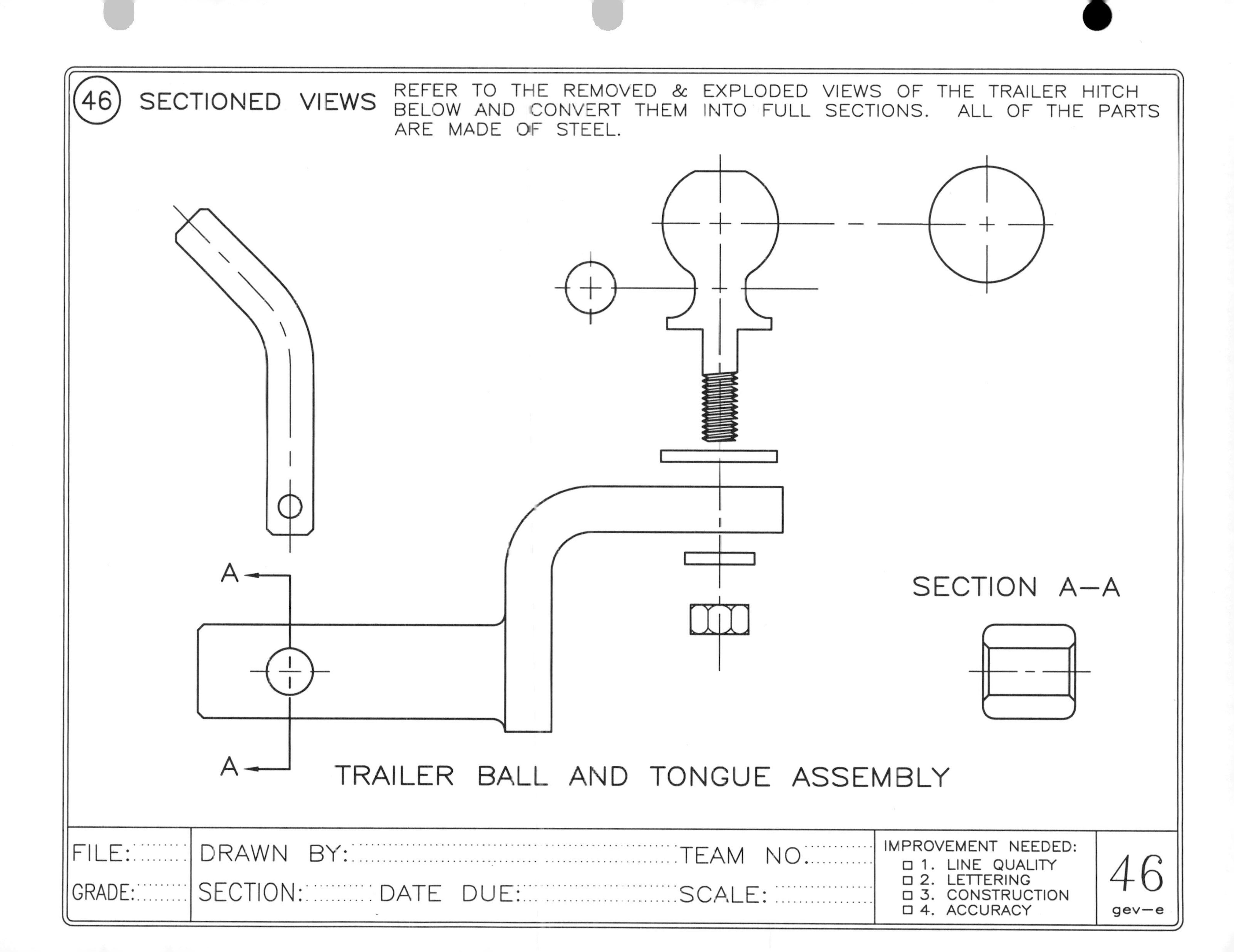
46
SECTIONED VIEWS
REFER TO THE REMOVED & EXPLODED VIEWS OF THE TRAILER HITCH BELOW AND CONVERT THEM INTO FULL SECTIONS. ALL OF THE PARTS ARE MADE OF STEEL.
A
A
SECTION A—A
TRAILER BALL AND TONGUE ASSEMBLY
FILE:
GRADE:
DRAWN BY:
TEAM NO.
SECTION:
DATE DUE:
SCALE:
IMPROVEMENT NEEDED:
1. LINE QUALITY
2. LETTERING
3. CONSTRUCTION
4. ACCURACY
46
gev—e

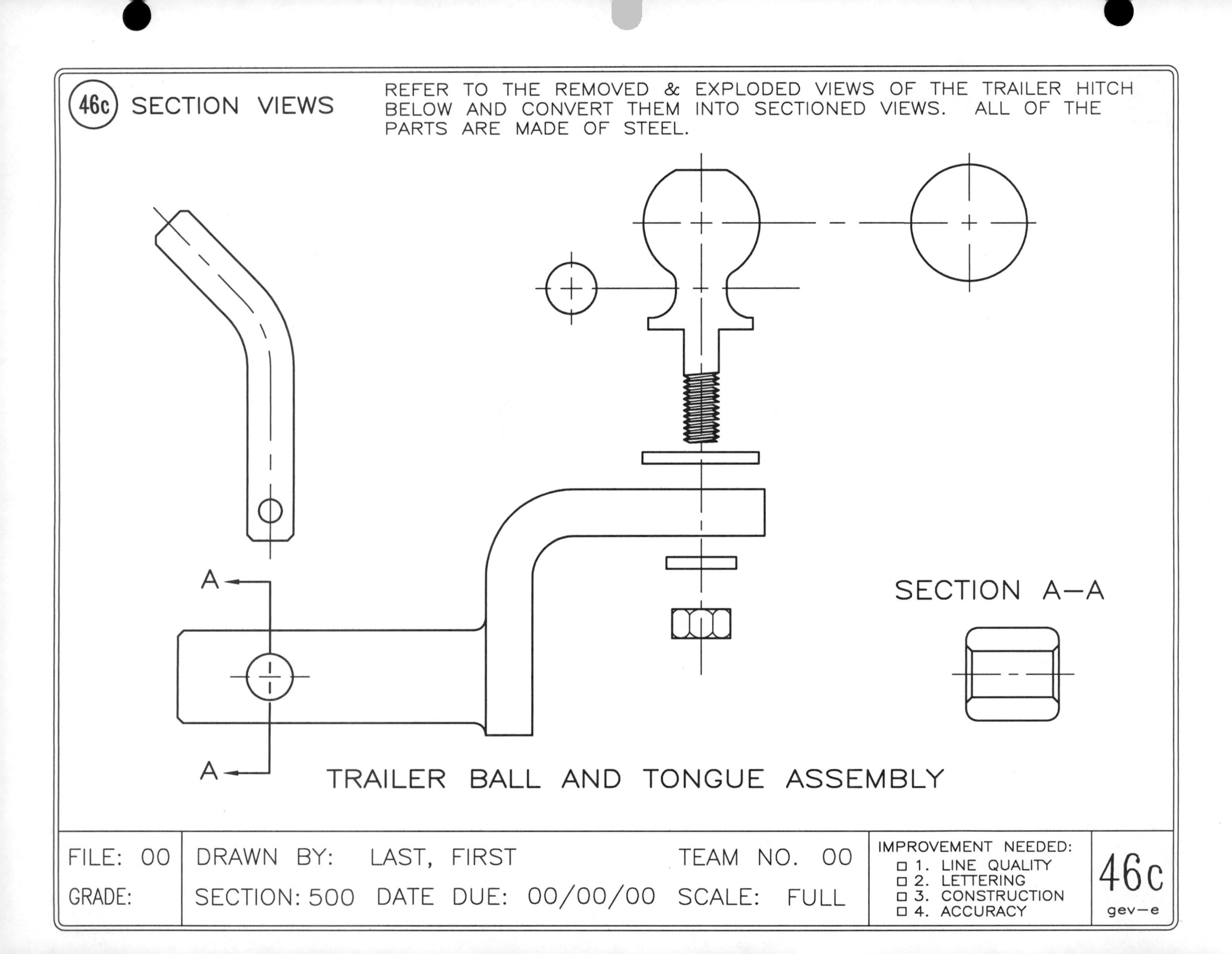
46c
SECTION VIEWS
REFER TO THE REMOVED & EXPLODED VIEWS OF THE TRAILER HITCH BELOW AND CONVERT THEM INTO SECTIONED VIEWS. ALL OF THE PARTS ARE MADE OF STEEL.
A
A
SECTION A—A
TRAILER BALL AND TONGUE ASSEMBLY
FILE: 00
GRADE:
DRAWN BY: LAST, FIRST
TEAM NO. 00
SECTION: 500 DATE DUE: 00/00/00 SCALE: FULL
IMPROVEMENT NEEDED:
☐ 1. LINE QUALITY
☐ 2. LETTERING
☐ 3. CONSTRUCTION
☐ 4. ACCURACY
46c
gev—e

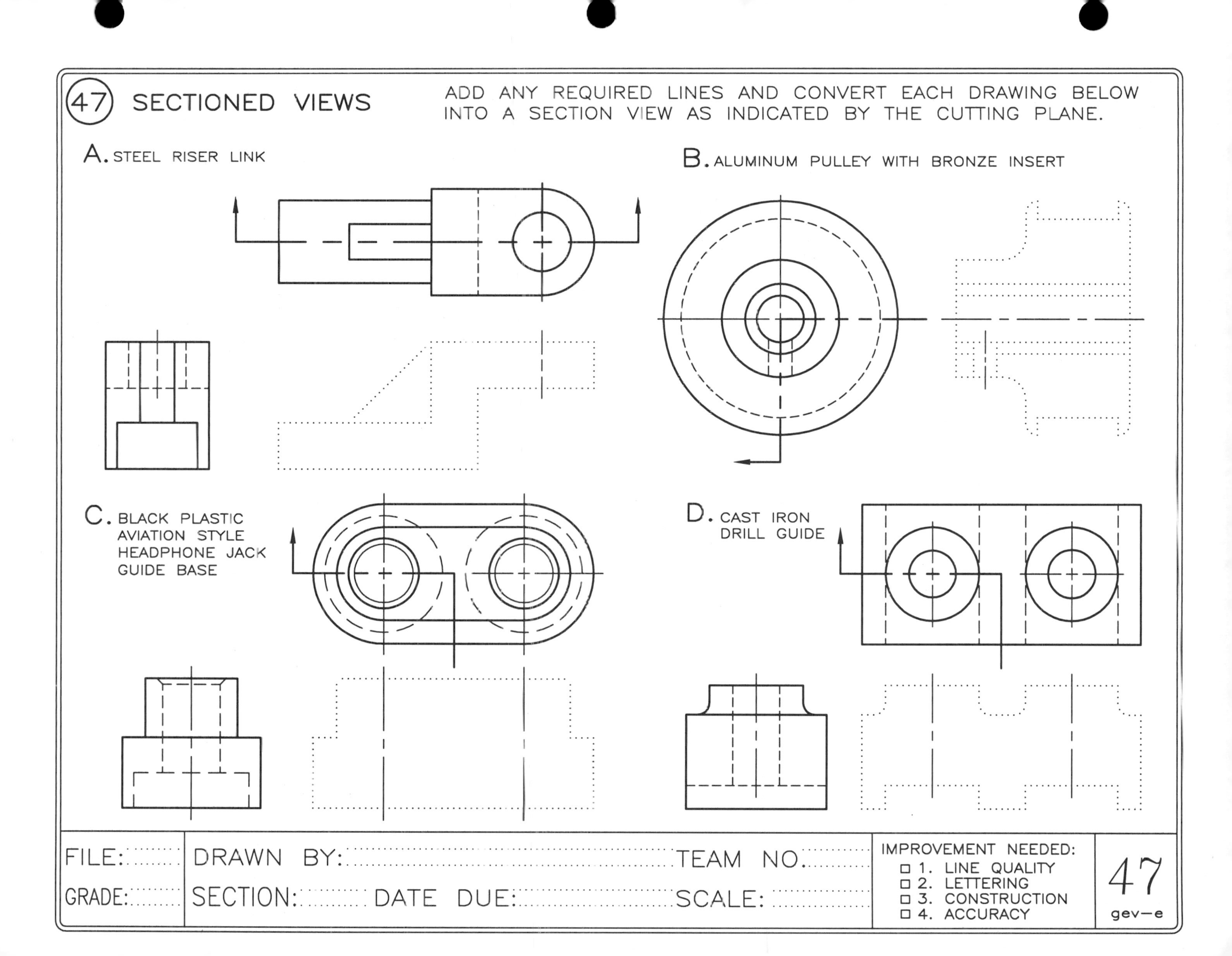
47
SECTIONED VIEWS
ADD ANY REQUIRED LINES AND CONVERT EACH DRAWING BELOW
INTO A SECTION VIEW AS INDICATED BY THE CUTTING PLANE.
A. STEEL RISER LINK
B. ALUMINUM PULLEY WITH BRONZE INSERT
C. BLACK PLASTIC
AVIATION STYLE
HEADPHONE JACK
GUIDE BASE
D. CAST IRON
DRILL GUIDE
FILE:
GRADE:
DRAWN BY:
TEAM NO.
SECTION:
DATE DUE:
SCALE:
IMPROVEMENT NEEDED:
☐ 1. LINE QUALITY
☐ 2. LETTERING
☐ 3. CONSTRUCTION
☐ 4. ACCURACY
47
gev-e

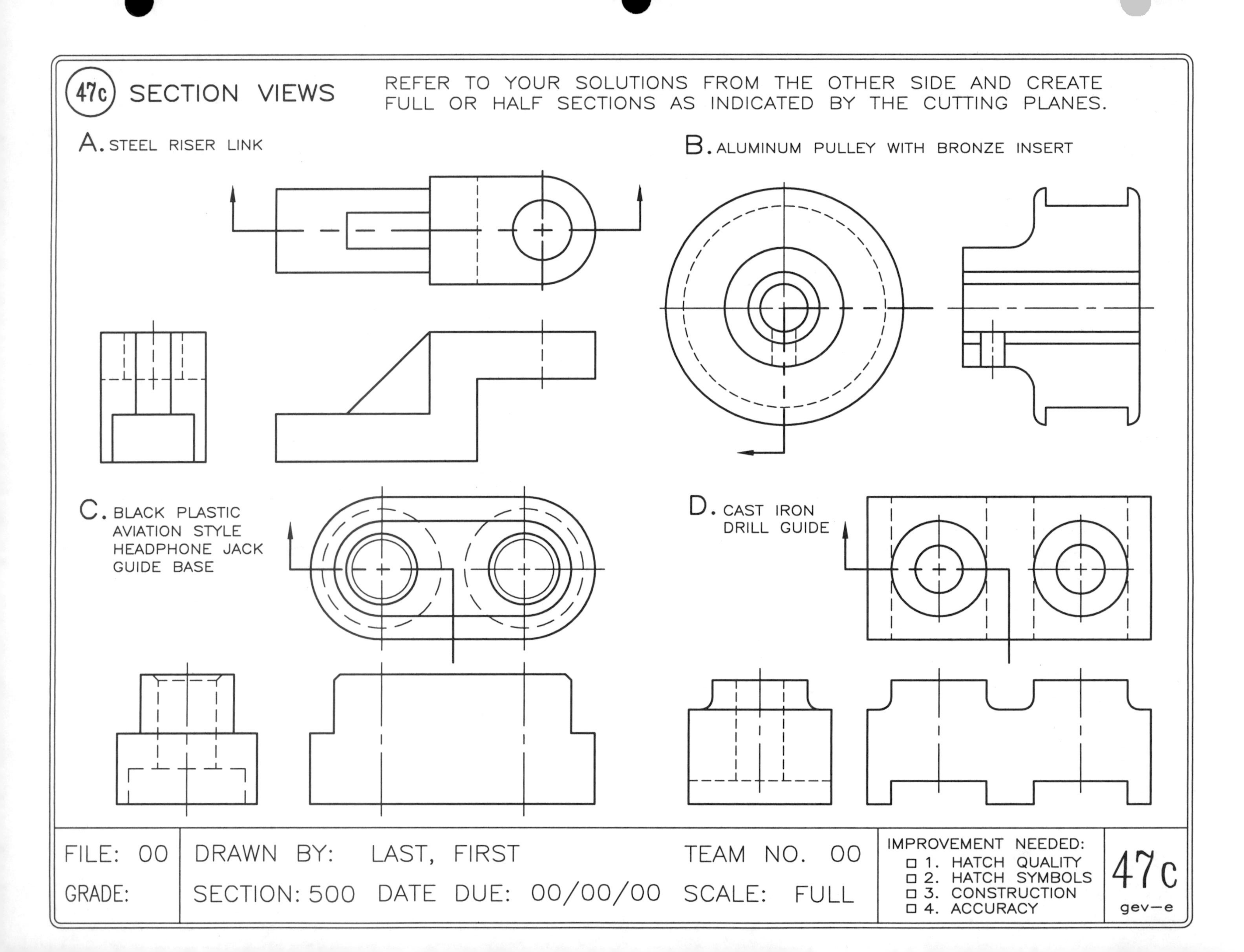
47c
SECTION VIEWS
REFER TO YOUR SOLUTIONS FROM THE OTHER SIDE AND CREATE FULL OR HALF SECTIONS AS INDICATED BY THE CUTTING PLANES.
A. STEEL RISER LINK
B. ALUMINUM PULLEY WITH BRONZE INSERT
C. BLACK PLASTIC AVIATION STYLE HEADPHONE JACK GUIDE BASE
D. CAST IRON DRILL GUIDE
FILE: 00
GRADE:
DRAWN BY: LAST, FIRST
SECTION: 500
DATE DUE: 00/00/00
TEAM NO. 00
SCALE: FULL
IMPROVEMENT NEEDED:
□ 1. HATCH QUALITY
□ 2. HATCH SYMBOLS
□ 3. CONSTRUCTION
□ 4. ACCURACY
47c
gev-e

(48) ASSORTED SECTIONS

COMPLETE EACH REVOLVED, REMOVED, OR BROKEN–OUT SECTION AS INDICATED BY THE CUTTING PLANES OR BREAK LINES.

A. CREATE REMOVED SECTIONS OF THE STEEL HAMMER HEAD.

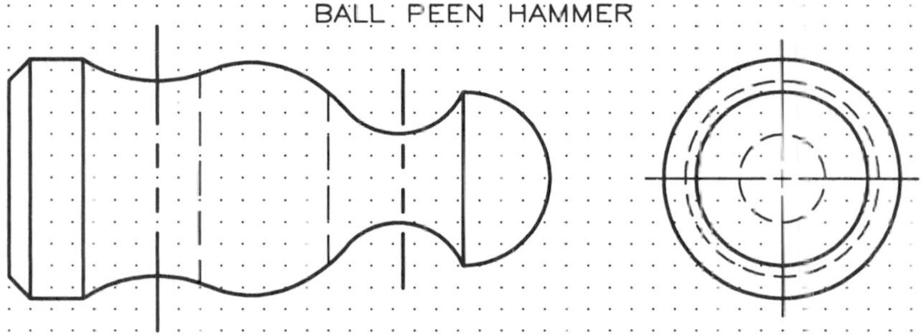

BALL PEEN HAMMER

B. CREATE A REVOLVED SECTION ON THE ROUND COPPER PIPE WHICH HAS CONVENTIONAL BREAKS ON EACH SIDE.

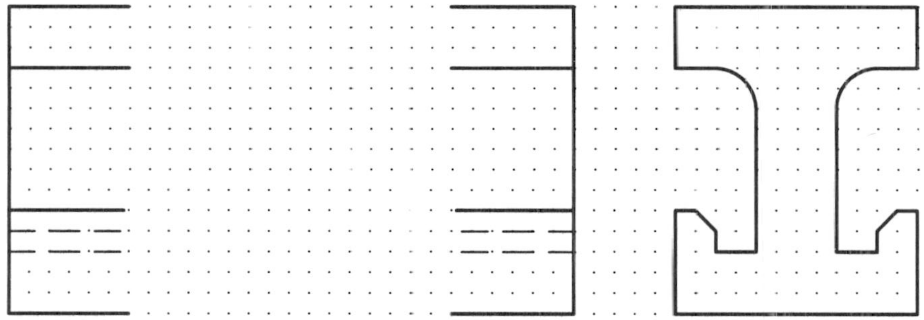

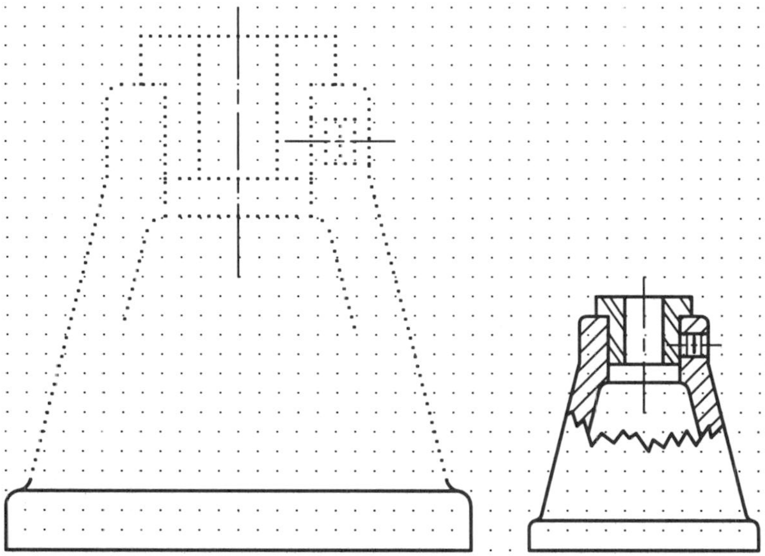

C. CREATE A REVOLVED SECTION ON THE OVERHEAD RAIL ABOVE WHICH HAS CONVENTIONAL BREAKS ON EACH SIDE.

D. CREATE A BROKEN–OUT SECTION OF THE BASE. EMPHASIZE CONTRAST BETWEEN THE PARTS.

FILE: .... DRAWN BY: .... TEAM NO. ....

GRADE: .... SECTION: .... DATE DUE: .... SCALE: ....

IMPROVEMENT NEEDED:

- ☐ 1. LINE QUALITY
- ☐ 2. LETTERING
- ☐ 3. CONSTRUCTION
- ☐ 4. ACCURACY

48

gev–e

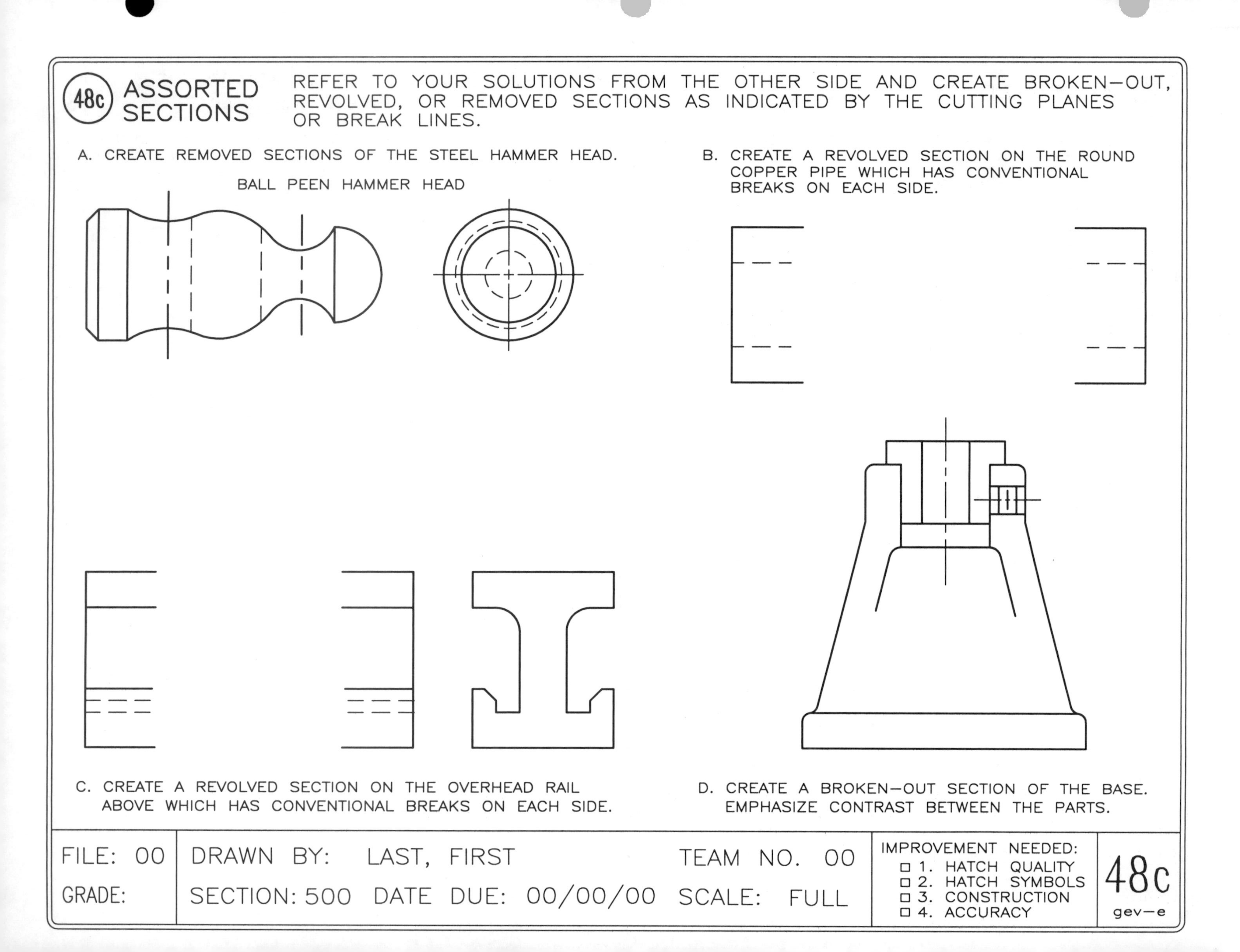
48c
ASSORTED SECTIONS
REFER TO YOUR SOLUTIONS FROM THE OTHER SIDE AND CREATE BROKEN—OUT, REVOLVED, OR REMOVED SECTIONS AS INDICATED BY THE CUTTING PLANES OR BREAK LINES.
A. CREATE REMOVED SECTIONS OF THE STEEL HAMMER HEAD.
BALL PEEN HAMMER HEAD
B. CREATE A REVOLVED SECTION ON THE ROUND COPPER PIPE WHICH HAS CONVENTIONAL BREAKS ON EACH SIDE.
C. CREATE A REVOLVED SECTION ON THE OVERHEAD RAIL ABOVE WHICH HAS CONVENTIONAL BREAKS ON EACH SIDE.
D. CREATE A BROKEN—OUT SECTION OF THE BASE. EMPHASIZE CONTRAST BETWEEN THE PARTS.
FILE: 00
GRADE:
DRAWN BY: LAST, FIRST
TEAM NO. 00
SECTION: 500 DATE DUE: 00/00/00 SCALE: FULL
IMPROVEMENT NEEDED:
□ 1. HATCH QUALITY
□ 2. HATCH SYMBOLS
□ 3. CONSTRUCTION
□ 4. ACCURACY
48c
gev—e

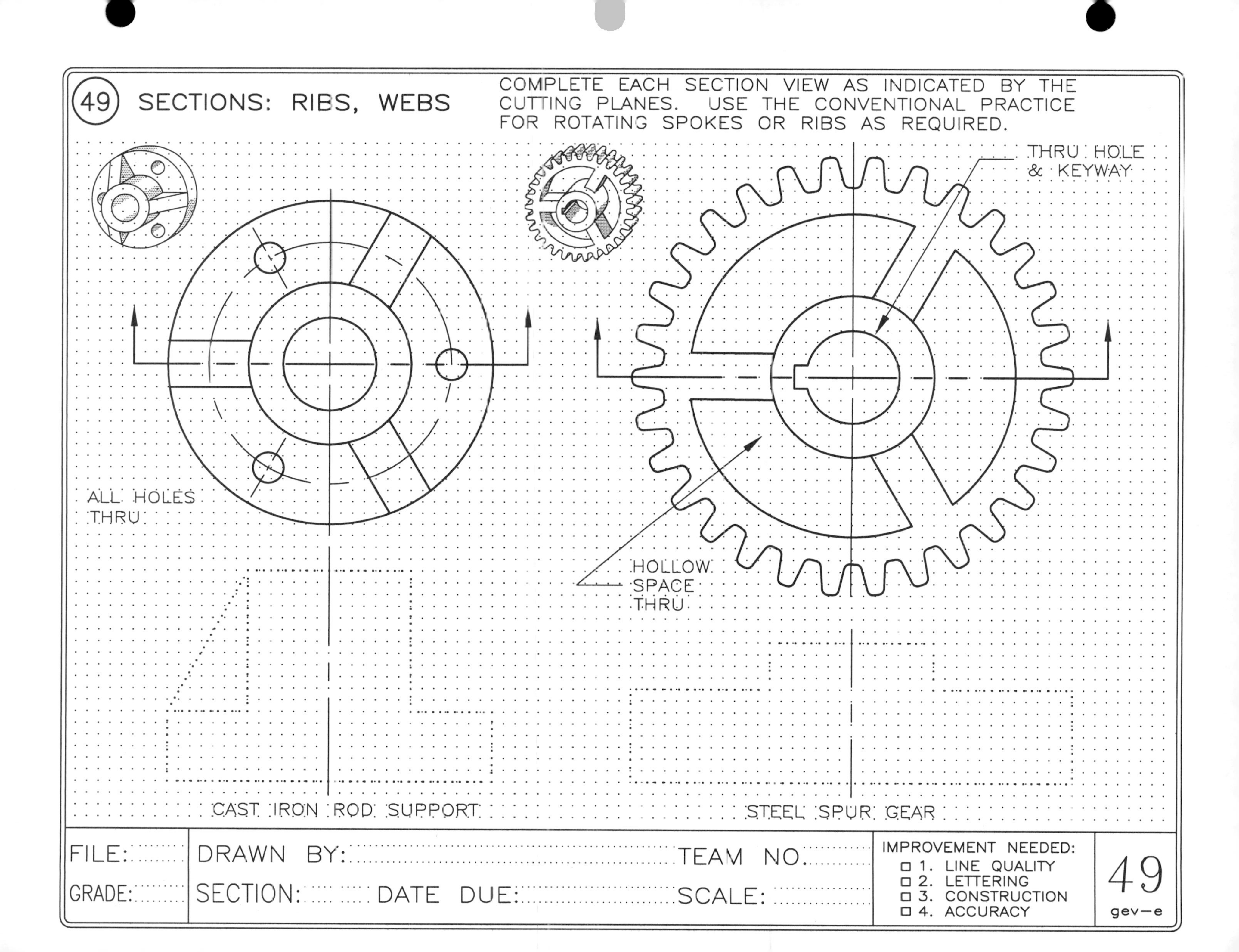
49
SECTIONS: RIBS, WEBS
COMPLETE EACH SECTION VIEW AS INDICATED BY THE CUTTING PLANES. USE THE CONVENTIONAL PRACTICE FOR ROTATING SPOKES OR RIBS AS REQUIRED.
THRU HOLE & KEYWAY
ALL HOLES THRU
HOLLOW SPACE THRU
CAST IRON ROD SUPPORT
STEEL SPUR GEAR
FILE:
GRADE:
DRAWN BY:
TEAM NO.
SECTION:
DATE DUE:
SCALE:
IMPROVEMENT NEEDED:
□ 1. LINE QUALITY
□ 2. LETTERING
□ 3. CONSTRUCTION
□ 4. ACCURACY
49
gev-e

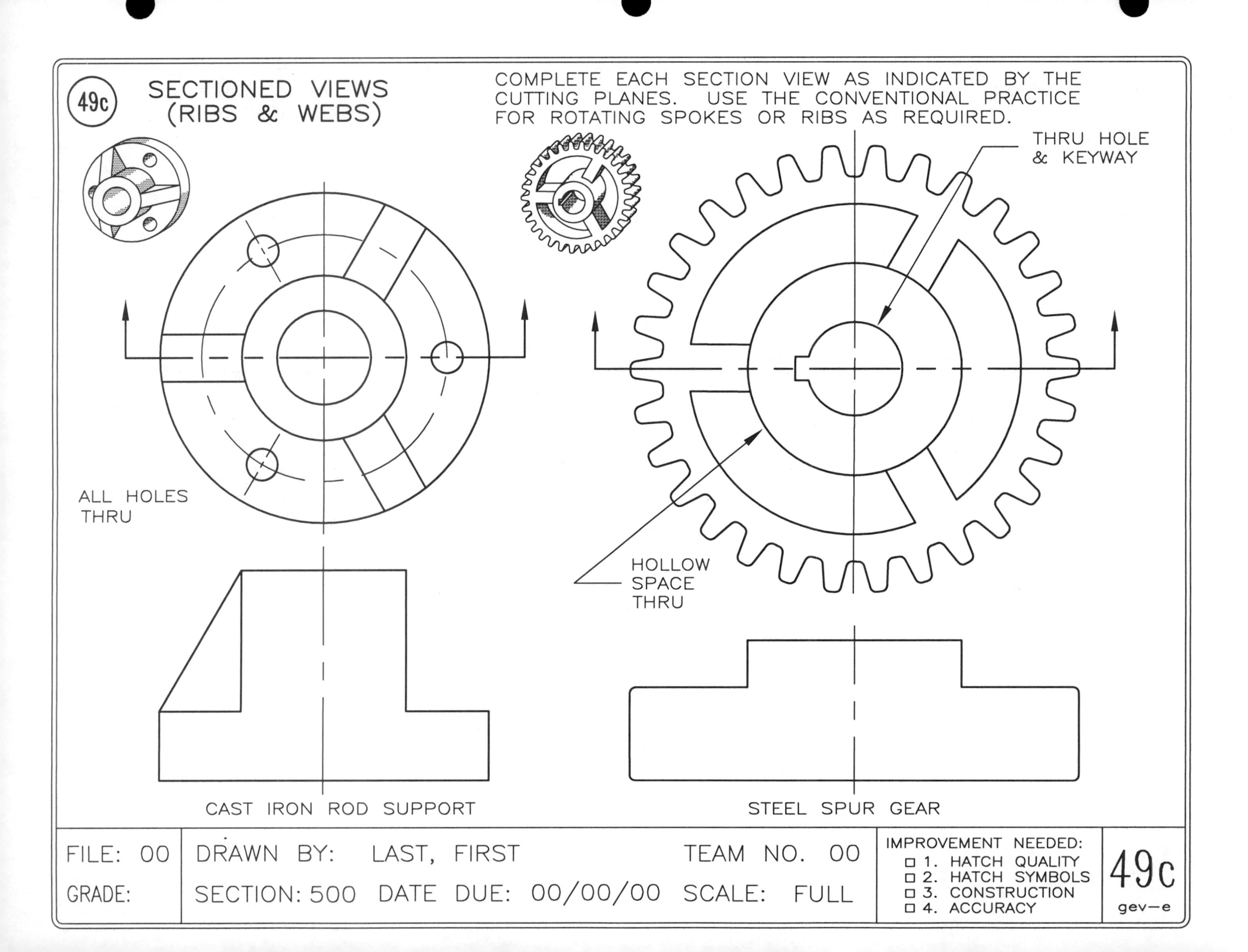
49c
SECTIONED VIEWS
(RIBS & WEBS)
COMPLETE EACH SECTION VIEW AS INDICATED BY THE CUTTING PLANES. USE THE CONVENTIONAL PRACTICE FOR ROTATING SPOKES OR RIBS AS REQUIRED.
THRU HOLE & KEYWAY
ALL HOLES THRU
HOLLOW SPACE THRU
CAST IRON ROD SUPPORT
STEEL SPUR GEAR
FILE: 00
GRADE:
DRAWN BY: LAST, FIRST
TEAM NO. 00
SECTION: 500 DATE DUE: 00/00/00 SCALE: FULL
IMPROVEMENT NEEDED:
□ 1. HATCH QUALITY
□ 2. HATCH SYMBOLS
□ 3. CONSTRUCTION
□ 4. ACCURACY
49c
gev-e

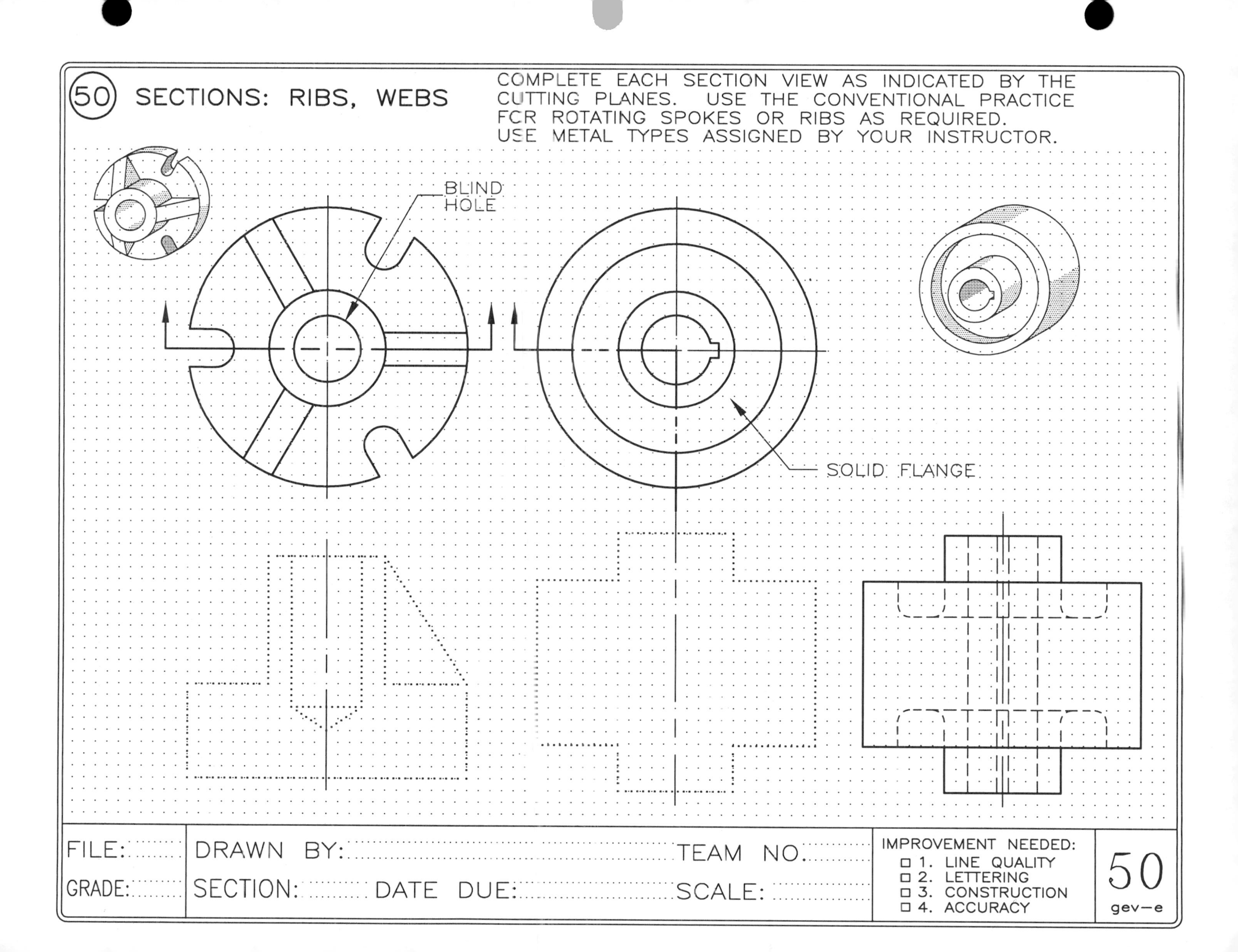
50
SECTIONS: RIBS, WEBS
COMPLETE EACH SECTION VIEW AS INDICATED BY THE CUTTING PLANES. USE THE CONVENTIONAL PRACTICE FOR ROTATING SPOKES OR RIBS AS REQUIRED.
USE METAL TYPES ASSIGNED BY YOUR INSTRUCTOR.
BLIND HOLE
SOLID FLANGE
FILE:
GRADE:
DRAWN BY:
TEAM NO.
SECTION:
DATE DUE:
SCALE:
IMPROVEMENT NEEDED:
1. LINE QUALITY
2. LETTERING
3. CONSTRUCTION
4. ACCURACY
50
gev–e

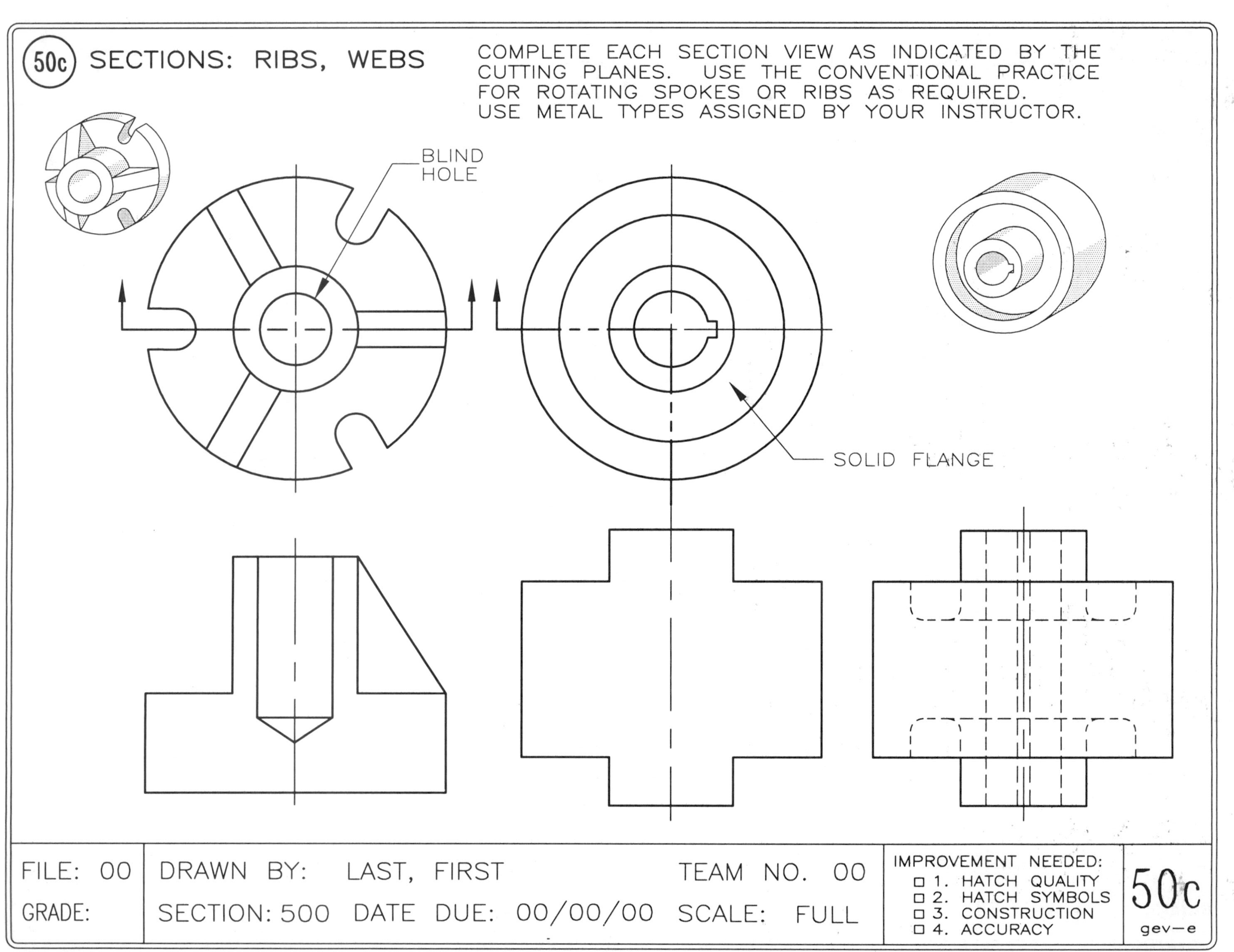
50c
SECTIONS: RIBS, WEBS
COMPLETE EACH SECTION VIEW AS INDICATED BY THE CUTTING PLANES. USE THE CONVENTIONAL PRACTICE FOR ROTATING SPOKES OR RIBS AS REQUIRED. USE METAL TYPES ASSIGNED BY YOUR INSTRUCTOR.
BLIND HOLE
SOLID FLANGE
FILE: 00
GRADE:
DRAWN BY: LAST, FIRST
TEAM NO. 00
SECTION: 500
DATE DUE: 00/00/00
SCALE: FULL
IMPROVEMENT NEEDED:
□ 1. HATCH QUALITY
□ 2. HATCH SYMBOLS
□ 3. CONSTRUCTION
□ 4. ACCURACY
50c
gev-e

51 STANDARD FEATURES

MODIFY EACH PART BELOW TO INCLUDE THE STANDARD FEATURES THAT ARE REQUESTED.

A. FILLETS

COMPLETE THE VIEW AND ADD .25"R FILLETS TO THE INSIDE CORNERS *

B. ROUNDS

COMPLETE THE VIEW AND ADD .25"R ROUNDS TO THE OUTSIDE CORNERS *

C. CHAMFERS

COMPLETE THE CYLINDER & ADD .25"x.25" CHAMFERS TO THE TOP AND BOTTOM

D. KEYWAYS

COMPLETE THE HOLE AND ADD .20"W x.10"DP KEYWAYS WHERE INDICATED WITH *

E. SLOTS

COMPLETE THE VIEW AND REPLACE THE RECTANGLES WITH 2.0" xØ.40" SLOTS

F. NECKS

COMPLETE THE VIEW AND ADD Ø.20"x.10" DP NECKS WHERE INDICATED WITH *

FILE:

GRADE:

DRAWN BY: TEAM NO.

SECTION: DATE DUE: SCALE:

IMPROVEMENT NEEDED:
- ☐ 1. LINE QUALITY
- ☐ 2. LETTERING
- ☐ 3. CONSTRUCTION
- ☐ 4. ACCURACY

51

gev-e

51c

STANDARD FEATURES

REFER TO YOUR SOLUTIONS ON THE OTHER SIDE OF THIS PAGE AND USE CAD TO MODIFY EACH PART BELOW WITH THE STANDARD FEATURES THAT ARE REQUESTED.

A. FILLETS

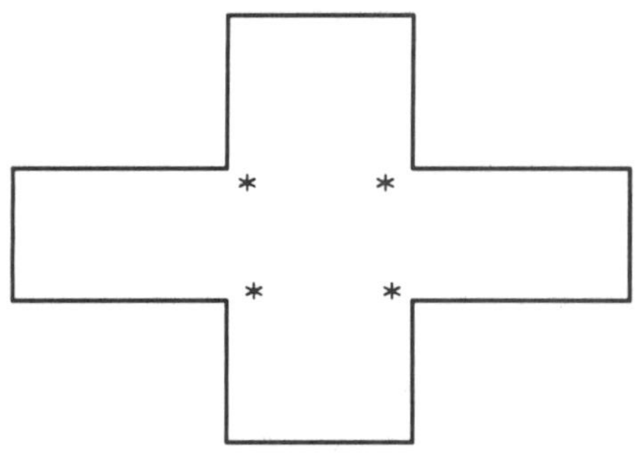

ADD .25"R FILLETS TO THE INSIDE CORNERS *

B. ROUNDS

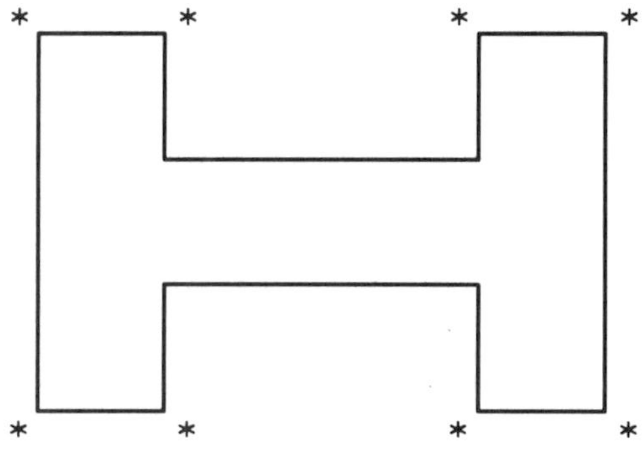

ADD .25"R ROUNDS TO THE OUTSIDE CORNERS *

C. CHAMFERS

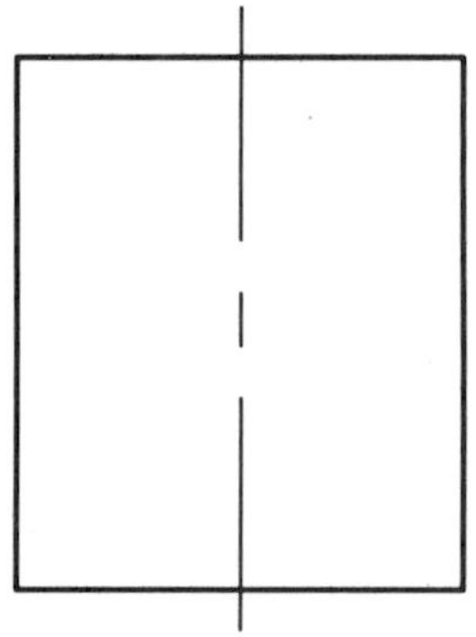

ADD .25"x.25" CHAMFERS TO THE TOP AND BOTTOM

D. KEYWAYS

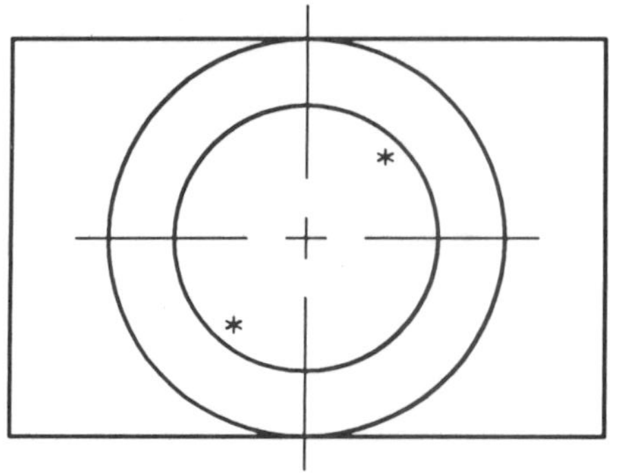

ADD .20"W x.10"DP KEYWAYS WHERE INDICATED WITH *

E. SLOTS

REPLACE THE RECTANGLES WITH 2.0"x Ø.40" SLOTS

F. NECKS

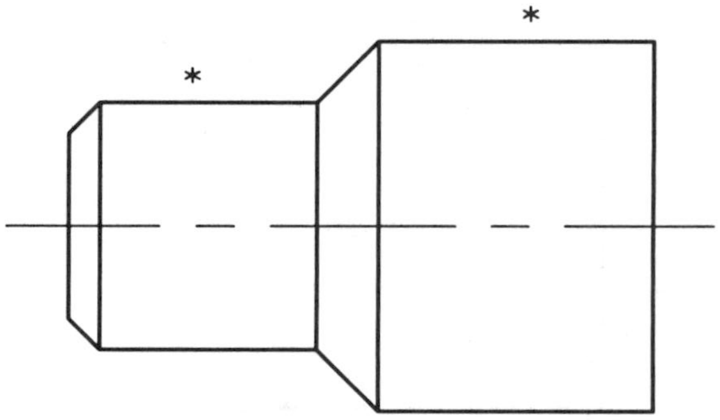

ADD Ø.20"x.10" DP NECKS WHERE INDICATED WITH *

FILE: 00	DRAWN BY: LAST, FIRST	TEAM NO. 00	IMPROVEMENT NEEDED:	51c
GRADE:	SECTION: 500 DATE DUE: 00/00/00	SCALE: FULL	☐ 1. FEATURES ☐ 2. LINE QUALITY ☐ 3. CONSTRUCTION ☐ 4. ACCURACY	gev—e

55 SCALE REVIEW

INDICATE THE SIZE OF EACH PART SHOWN BELOW IN THE SPACE PROVIDED. USE THE SCALE INDICATED OR ONE THAT IS ASSIGNED BY YOUR INSTRUCTOR.

1. SCALE: 1=10

A. ....
B. ....
C. ....
D. ....
E. ....

2. SCALE: 1=200

A. ....
B. ....
C. ....
D. ....
E. ....

3. SCALE: 1=4

A. ....
B. ....
C. ....
D. ....
E. ....

4. SCALE: 1:1 (METRIC)

L. ....
M. ....
N. ....
O. ....
P. ....

5. SCALE: 1:50 (METRIC)

L. ....
M. ....
N. ....
O. ....
P. ....

6. SCALE: 1:400 (METRIC)

L. ....
M. ....
N. ....
O. ....
P. ....

FILE: .... DRAWN BY: .... TEAM NO. ....

GRADE: .... SECTION: .... DATE DUE: .... SCALE: ....

IMPROVEMENT NEEDED:
- ☐ 1. LINE QUALITY
- ☐ 2. LETTERING
- ☐ 3. CONSTRUCTION
- ☐ 4. ACCURACY

55

gev—e

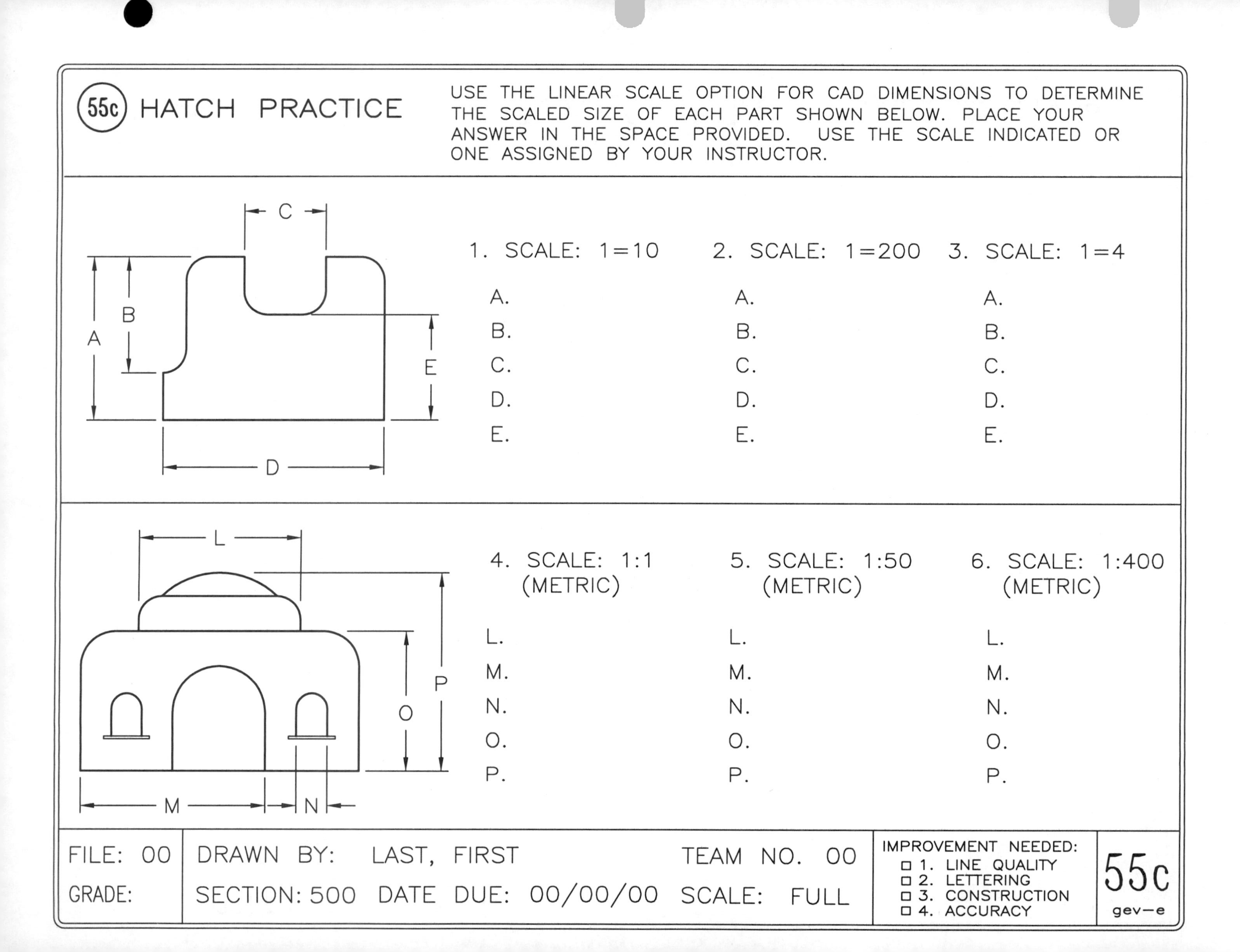

55c
HATCH PRACTICE
USE THE LINEAR SCALE OPTION FOR CAD DIMENSIONS TO DETERMINE THE SCALED SIZE OF EACH PART SHOWN BELOW. PLACE YOUR ANSWER IN THE SPACE PROVIDED. USE THE SCALE INDICATED OR ONE ASSIGNED BY YOUR INSTRUCTOR.
C
B
A
E
D
1. SCALE: 1=10
A.
B.
C.
D.
E.
2. SCALE: 1=200
A.
B.
C.
D.
E.
3. SCALE: 1=4
A.
B.
C.
D.
E.
L
P
O
M
N
4. SCALE: 1:1 (METRIC)
L.
M.
N.
O.
P.
5. SCALE: 1:50 (METRIC)
L.
M.
N.
O.
P.
6. SCALE: 1:400 (METRIC)
L.
M.
N.
O.
P.
FILE: 00
GRADE:
DRAWN BY: LAST, FIRST
TEAM NO. 00
SECTION: 500 DATE DUE: 00/00/00 SCALE: FULL
IMPROVEMENT NEEDED:
□ 1. LINE QUALITY
□ 2. LETTERING
□ 3. CONSTRUCTION
□ 4. ACCURACY
55c
gev-e

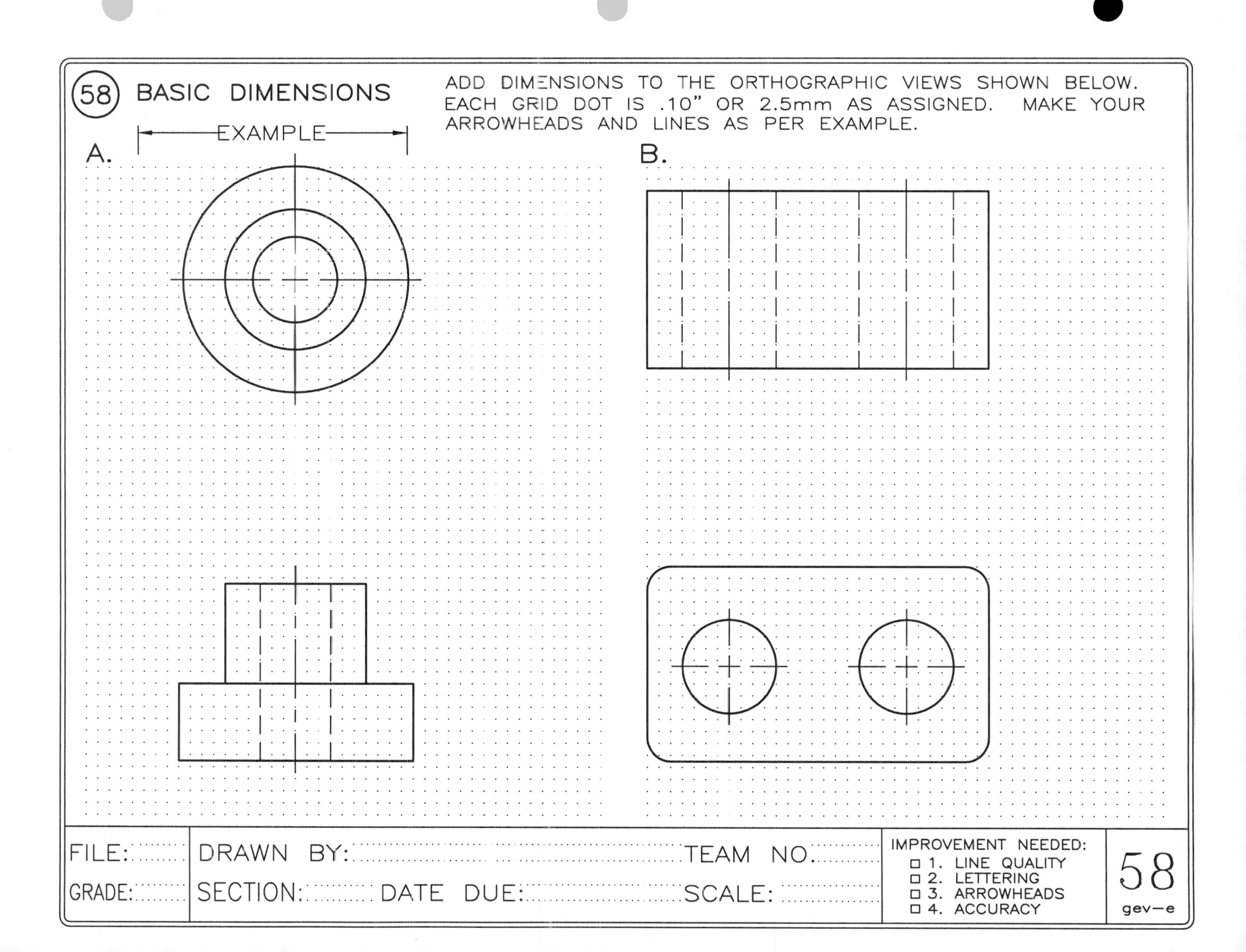
58
BASIC DIMENSIONS
ADD DIMENSIONS TO THE ORTHOGRAPHIC VIEWS SHOWN BELOW. EACH GRID DOT IS .10" OR 2.5mm AS ASSIGNED. MAKE YOUR ARROWHEADS AND LINES AS PER EXAMPLE.
EXAMPLE
A.
B.
FILE:
GRADE:
DRAWN BY:
TEAM NO.
SECTION:
DATE DUE:
SCALE:
IMPROVEMENT NEEDED:
1. LINE QUALITY
2. LETTERING
3. ARROWHEADS
4. ACCURACY
58
gev-e

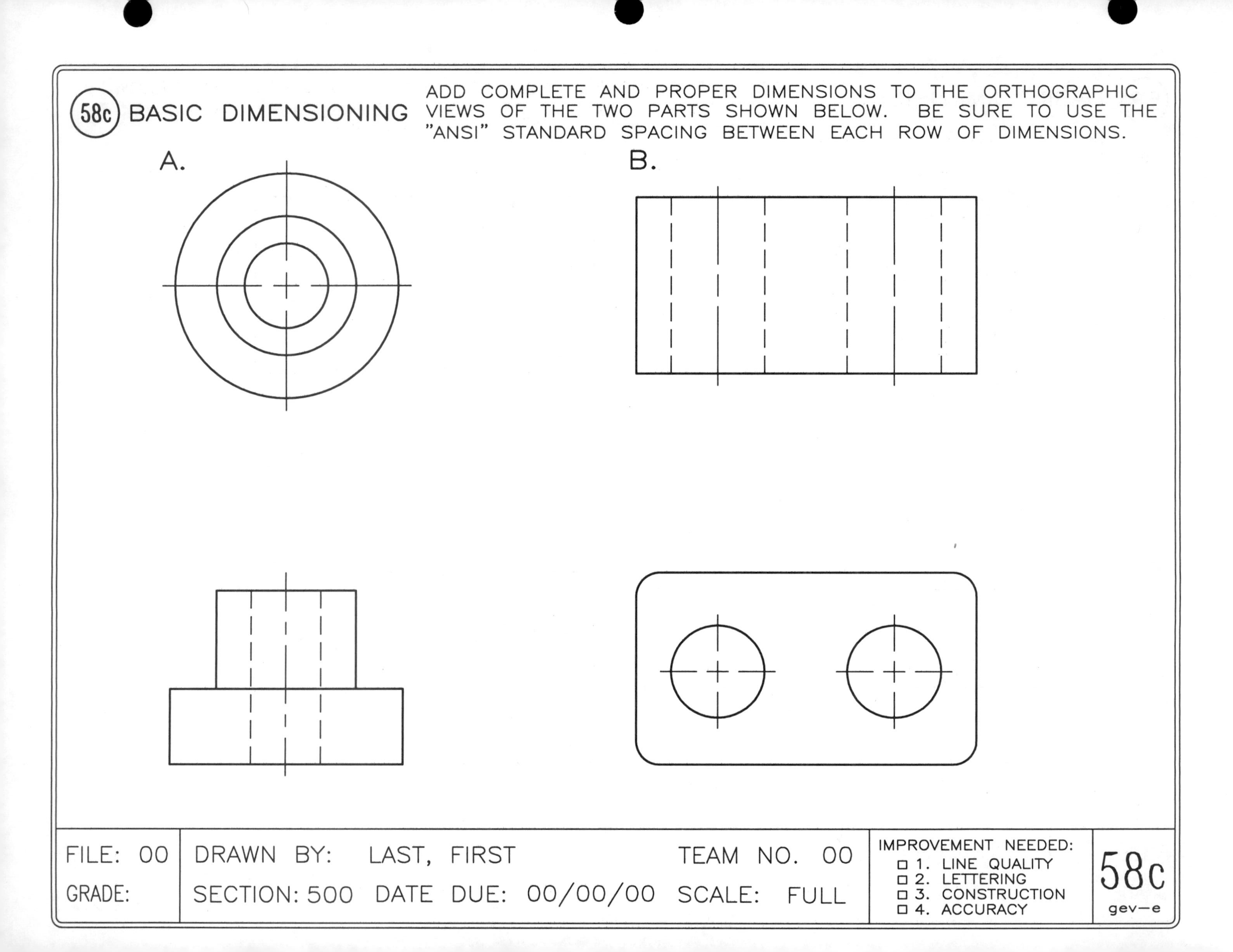
58c
BASIC DIMENSIONING
ADD COMPLETE AND PROPER DIMENSIONS TO THE ORTHOGRAPHIC VIEWS OF THE TWO PARTS SHOWN BELOW. BE SURE TO USE THE "ANSI" STANDARD SPACING BETWEEN EACH ROW OF DIMENSIONS.
A.
B.
FILE: 00
GRADE:
DRAWN BY: LAST, FIRST
TEAM NO. 00
SECTION: 500
DATE DUE: 00/00/00
SCALE: FULL
IMPROVEMENT NEEDED:
□ 1. LINE QUALITY
□ 2. LETTERING
□ 3. CONSTRUCTION
□ 4. ACCURACY
58c
gev—e

(59) BASIC DIMENSIONS

ADD DIMENSIONS TO THE ORTHOGRAPHIC VIEWS SHOWN BELOW. EACH GRID DOT IS .10" OR 2.5mm AS ASSIGNED. MAKE YOUR ARROWHEADS AND LINES AS PER EXAMPLE.

EXAMPLE

A.

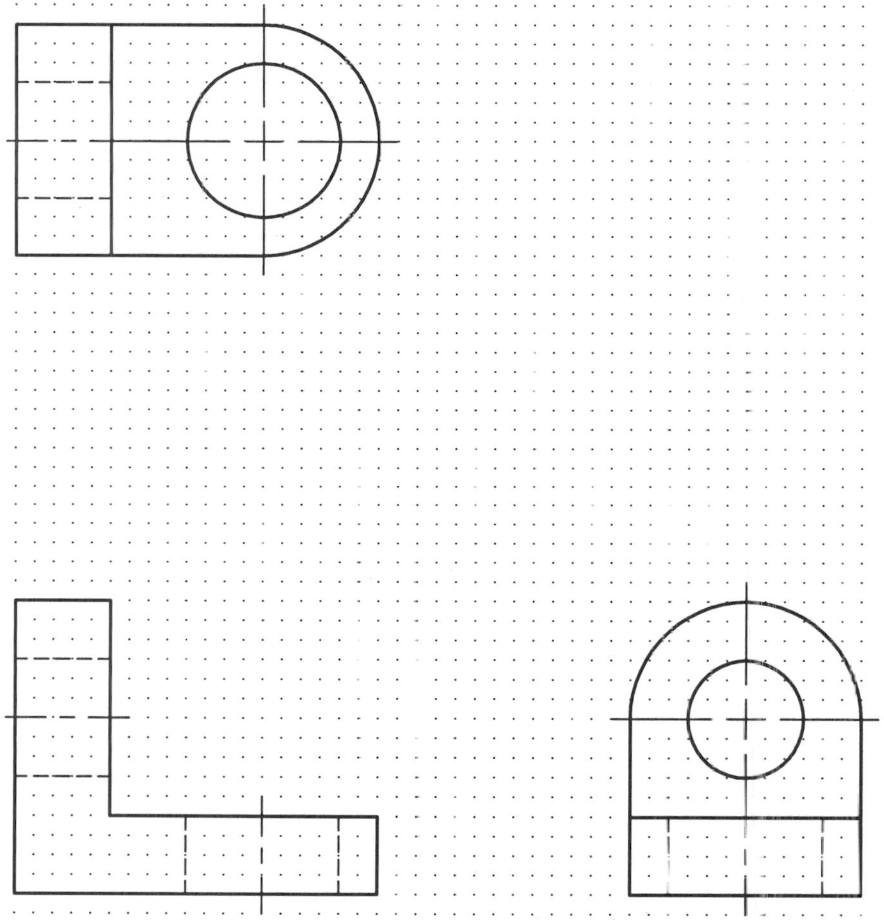

B.

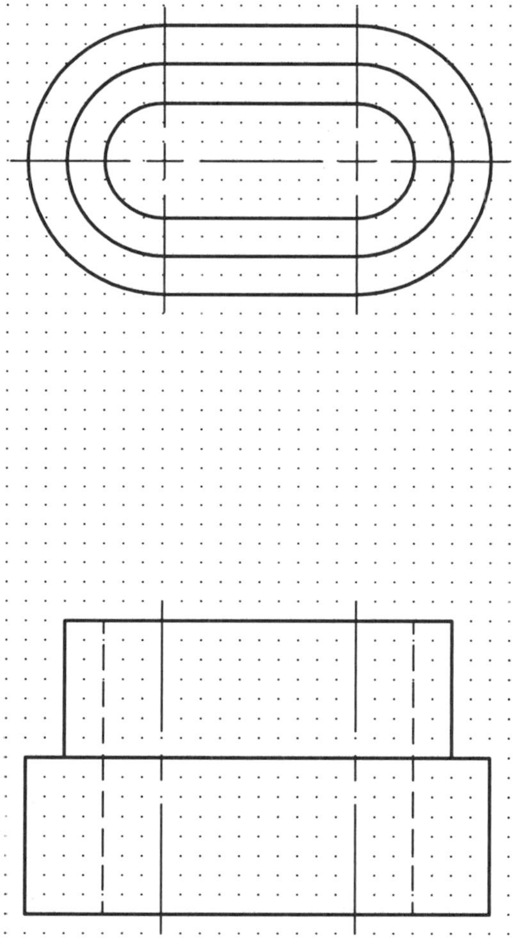

FILE:	DRAWN BY:	TEAM NO.	IMPROVEMENT NEEDED: ☐ 1. LINE QUALITY ☐ 2. LETTERING ☐ 3. ARROWHEADS ☐ 4. ACCURACY	59 gev–e
GRADE:	SECTION: DATE DUE:	SCALE:		

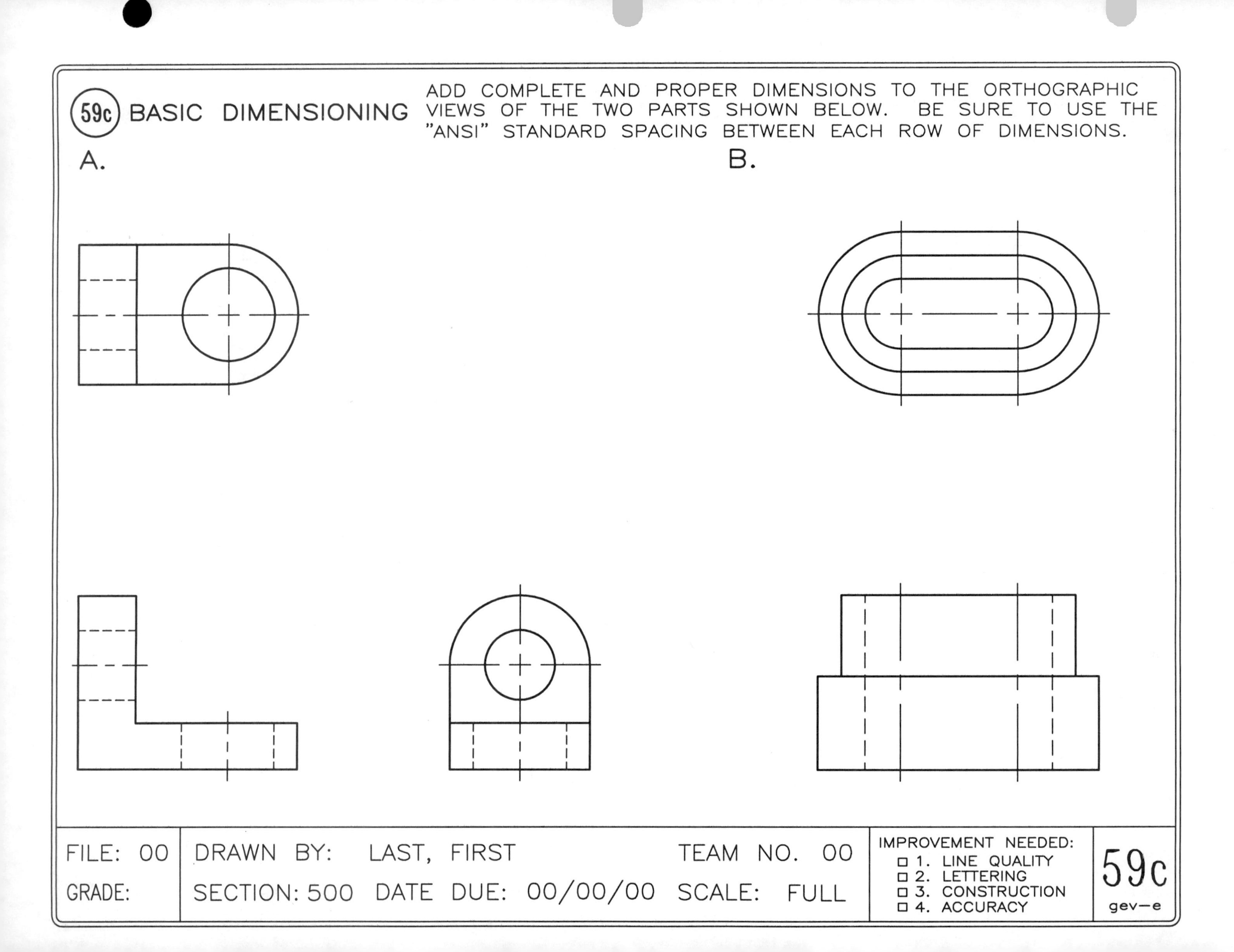
59c
BASIC DIMENSIONING
ADD COMPLETE AND PROPER DIMENSIONS TO THE ORTHOGRAPHIC VIEWS OF THE TWO PARTS SHOWN BELOW. BE SURE TO USE THE "ANSI" STANDARD SPACING BETWEEN EACH ROW OF DIMENSIONS.
A.
B.
FILE: 00
GRADE:
DRAWN BY: LAST, FIRST
SECTION: 500
DATE DUE: 00/00/00
TEAM NO. 00
SCALE: FULL
IMPROVEMENT NEEDED:
☐ 1. LINE QUALITY
☐ 2. LETTERING
☐ 3. CONSTRUCTION
☐ 4. ACCURACY
59c
gev—e

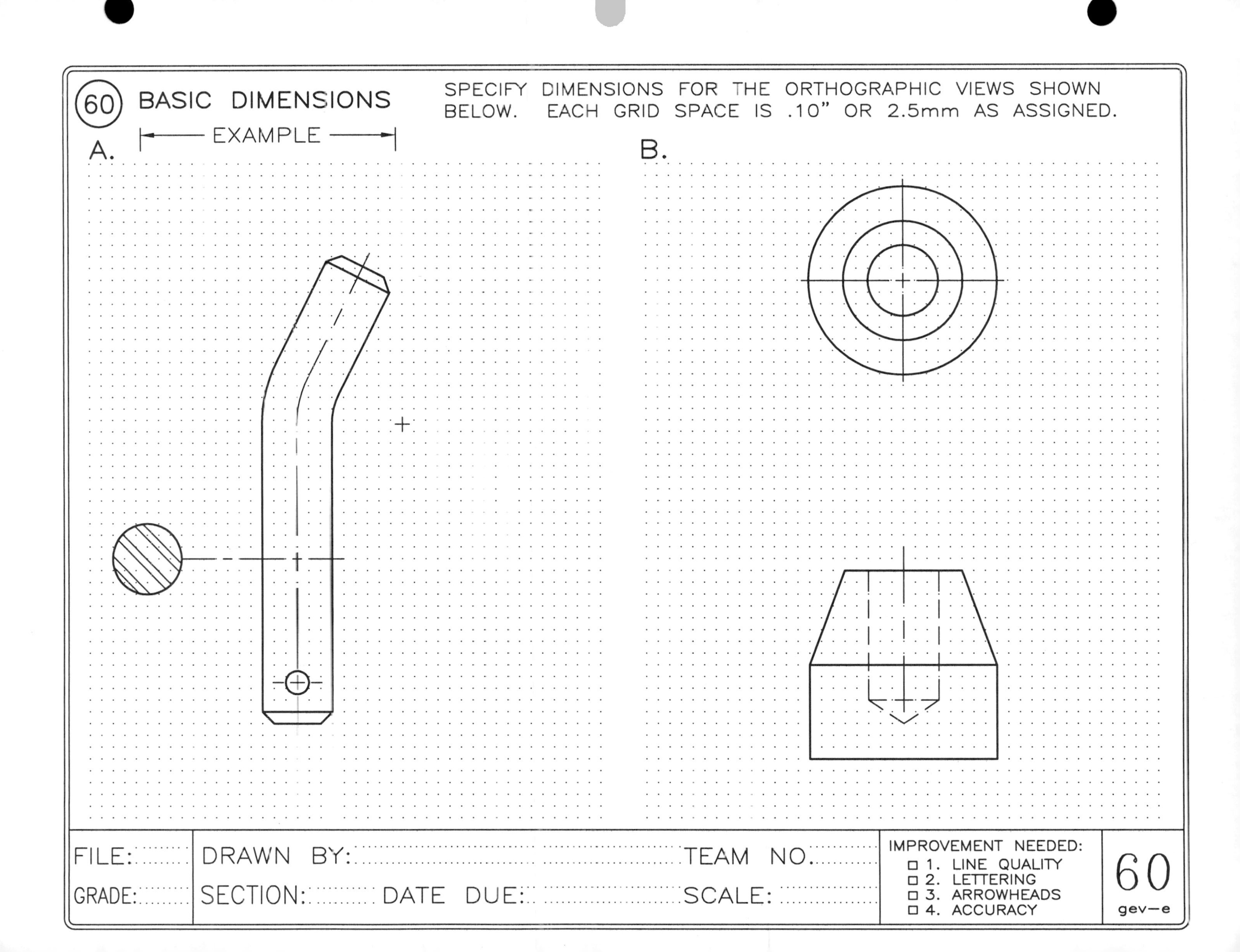
60
BASIC DIMENSIONS
SPECIFY DIMENSIONS FOR THE ORTHOGRAPHIC VIEWS SHOWN BELOW. EACH GRID SPACE IS .10" OR 2.5mm AS ASSIGNED.
EXAMPLE
A.
B.
FILE:
GRADE:
DRAWN BY:
TEAM NO.
SECTION:
DATE DUE:
SCALE:
IMPROVEMENT NEEDED:
☐ 1. LINE QUALITY
☐ 2. LETTERING
☐ 3. ARROWHEADS
☐ 4. ACCURACY
60
gev-e

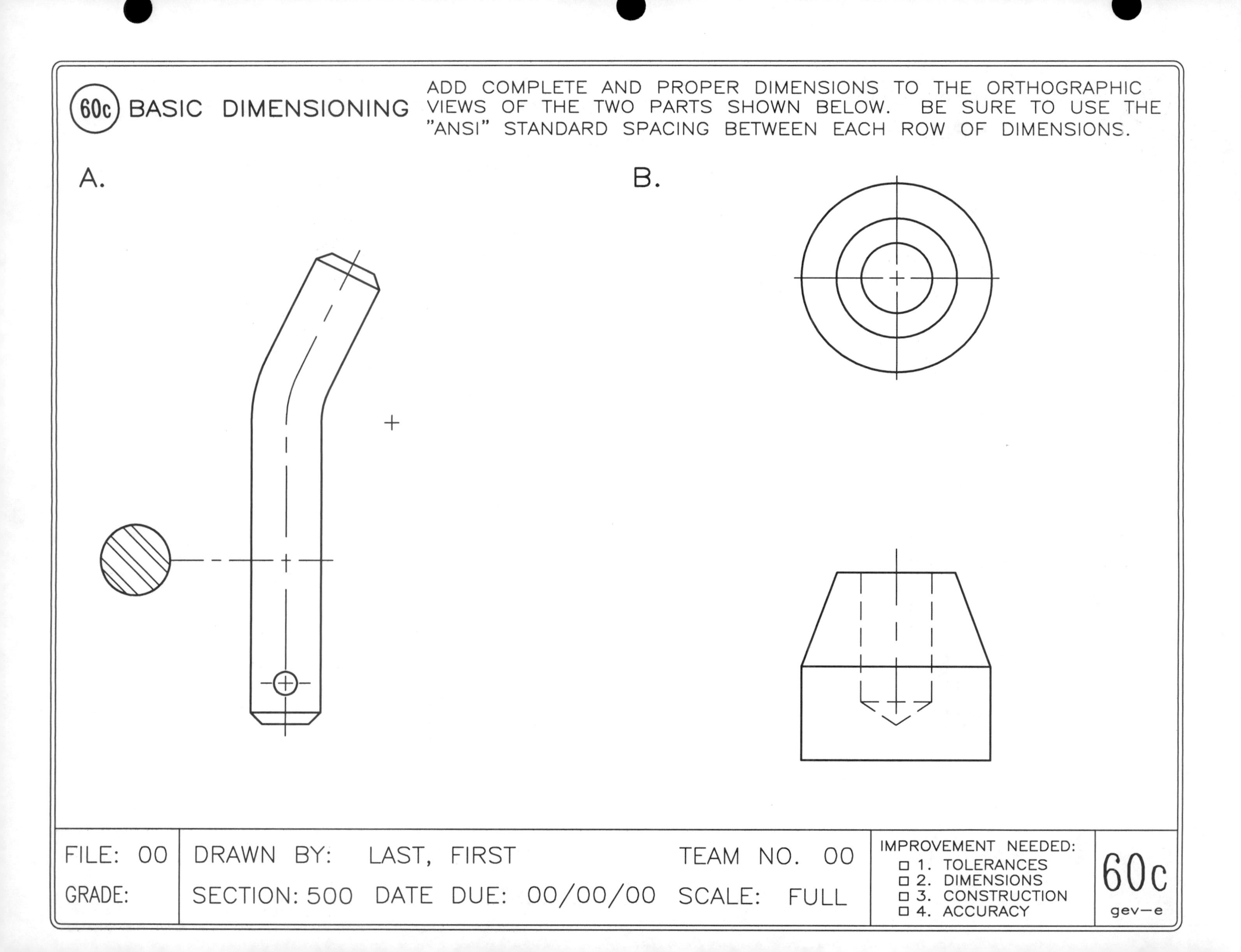
60c
BASIC DIMENSIONING
ADD COMPLETE AND PROPER DIMENSIONS TO THE ORTHOGRAPHIC VIEWS OF THE TWO PARTS SHOWN BELOW. BE SURE TO USE THE "ANSI" STANDARD SPACING BETWEEN EACH ROW OF DIMENSIONS.
A.
B.
FILE: 00
GRADE:
DRAWN BY: LAST, FIRST
TEAM NO. 00
SECTION: 500 DATE DUE: 00/00/00 SCALE: FULL
IMPROVEMENT NEEDED:
☐ 1. TOLERANCES
☐ 2. DIMENSIONS
☐ 3. CONSTRUCTION
☐ 4. ACCURACY
60c
gev—e

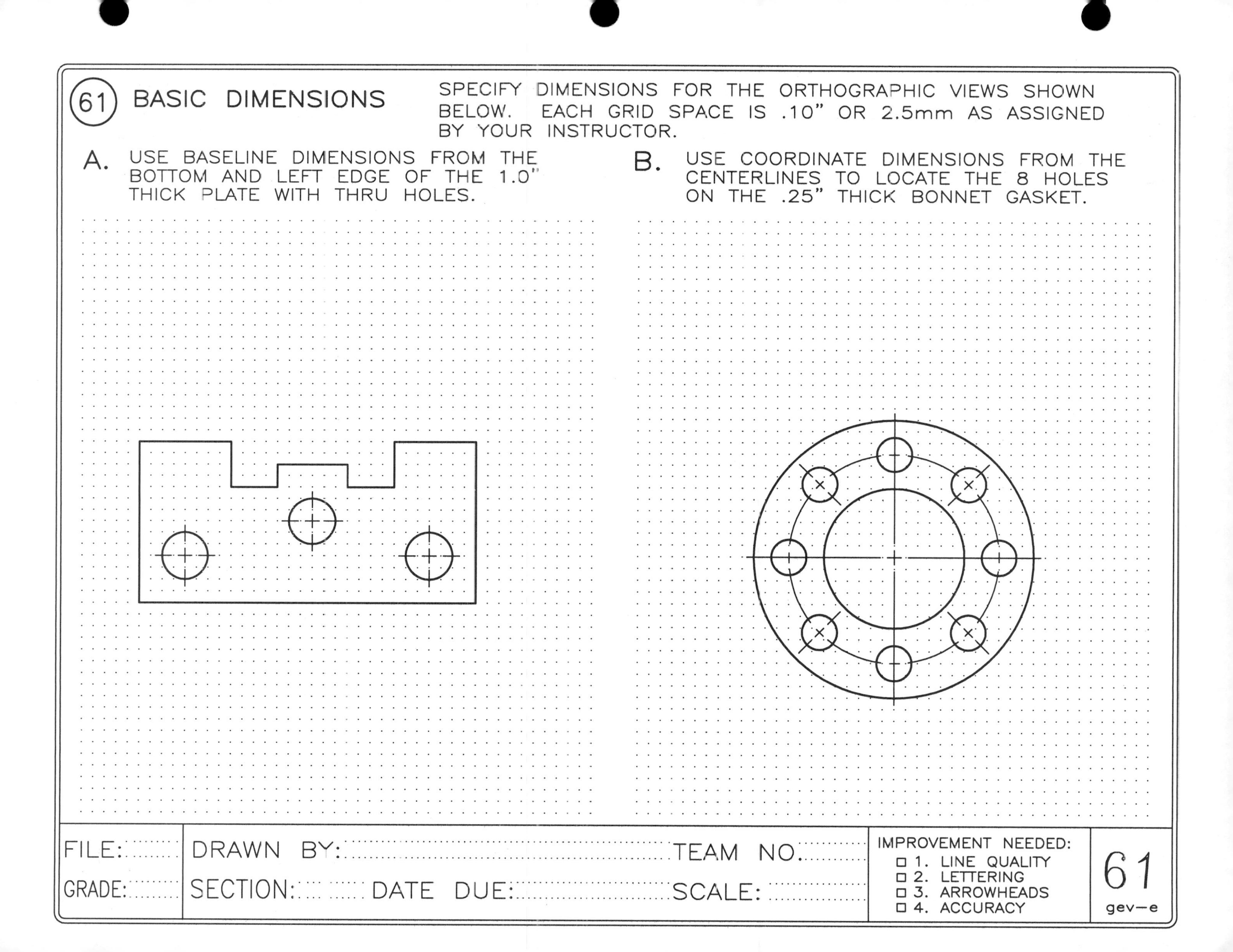
61
BASIC DIMENSIONS
SPECIFY DIMENSIONS FOR THE ORTHOGRAPHIC VIEWS SHOWN BELOW. EACH GRID SPACE IS .10" OR 2.5mm AS ASSIGNED BY YOUR INSTRUCTOR.
A. USE BASELINE DIMENSIONS FROM THE BOTTOM AND LEFT EDGE OF THE 1.0" THICK PLATE WITH THRU HOLES.
B. USE COORDINATE DIMENSIONS FROM THE CENTERLINES TO LOCATE THE 8 HOLES ON THE .25" THICK BONNET GASKET.
FILE:
GRADE:
DRAWN BY:
TEAM NO.
SECTION:
DATE DUE:
SCALE:
IMPROVEMENT NEEDED:
□ 1. LINE QUALITY
□ 2. LETTERING
□ 3. ARROWHEADS
□ 4. ACCURACY
61
gev–e

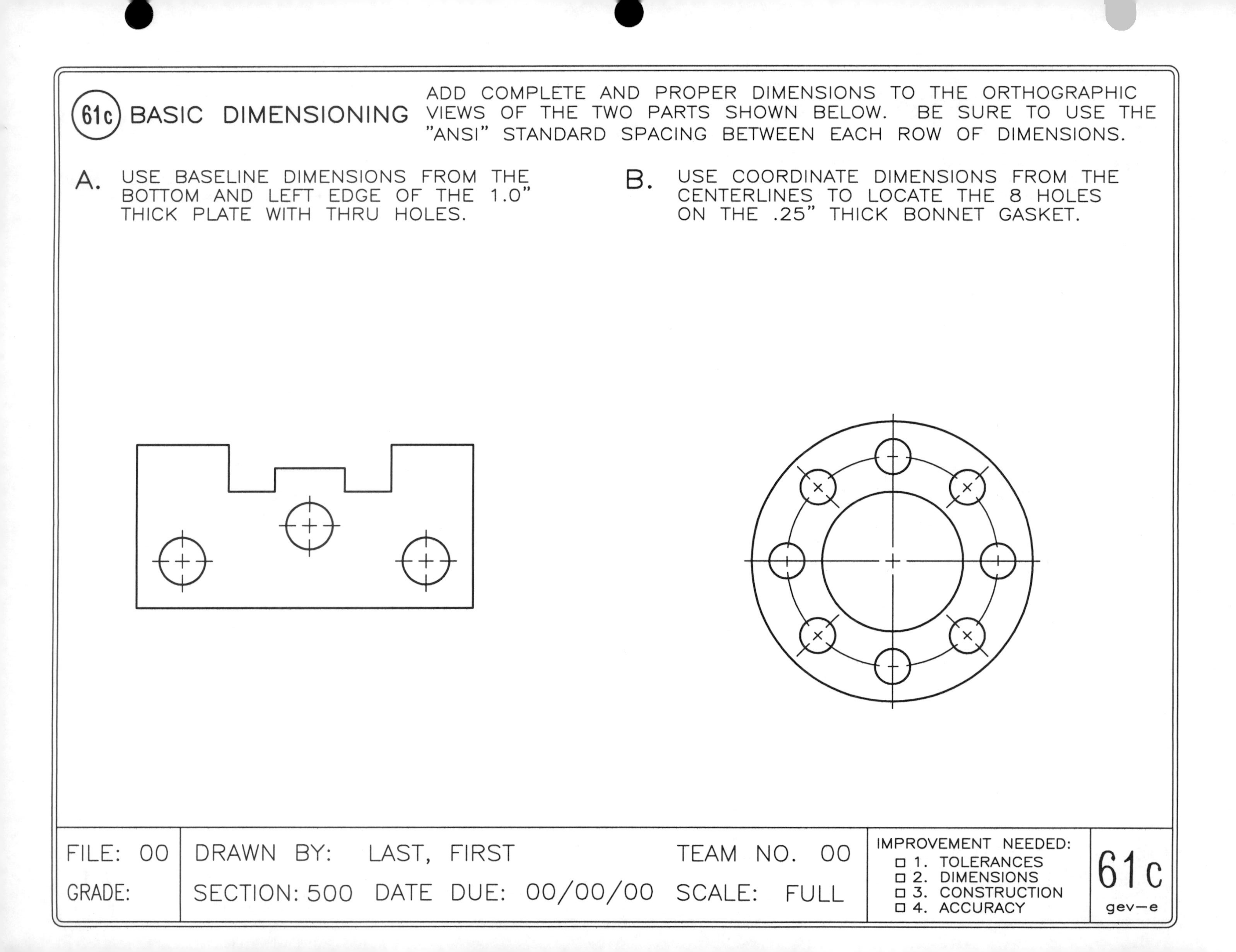
61c
BASIC DIMENSIONING
ADD COMPLETE AND PROPER DIMENSIONS TO THE ORTHOGRAPHIC VIEWS OF THE TWO PARTS SHOWN BELOW. BE SURE TO USE THE "ANSI" STANDARD SPACING BETWEEN EACH ROW OF DIMENSIONS.
A. USE BASELINE DIMENSIONS FROM THE BOTTOM AND LEFT EDGE OF THE 1.0" THICK PLATE WITH THRU HOLES.
B. USE COORDINATE DIMENSIONS FROM THE CENTERLINES TO LOCATE THE 8 HOLES ON THE .25" THICK BONNET GASKET.
FILE: 00
GRADE:
DRAWN BY: LAST, FIRST
TEAM NO. 00
SECTION: 500
DATE DUE: 00/00/00
SCALE: FULL
IMPROVEMENT NEEDED:
☐ 1. TOLERANCES
☐ 2. DIMENSIONS
☐ 3. CONSTRUCTION
☐ 4. ACCURACY
61c
gev-e

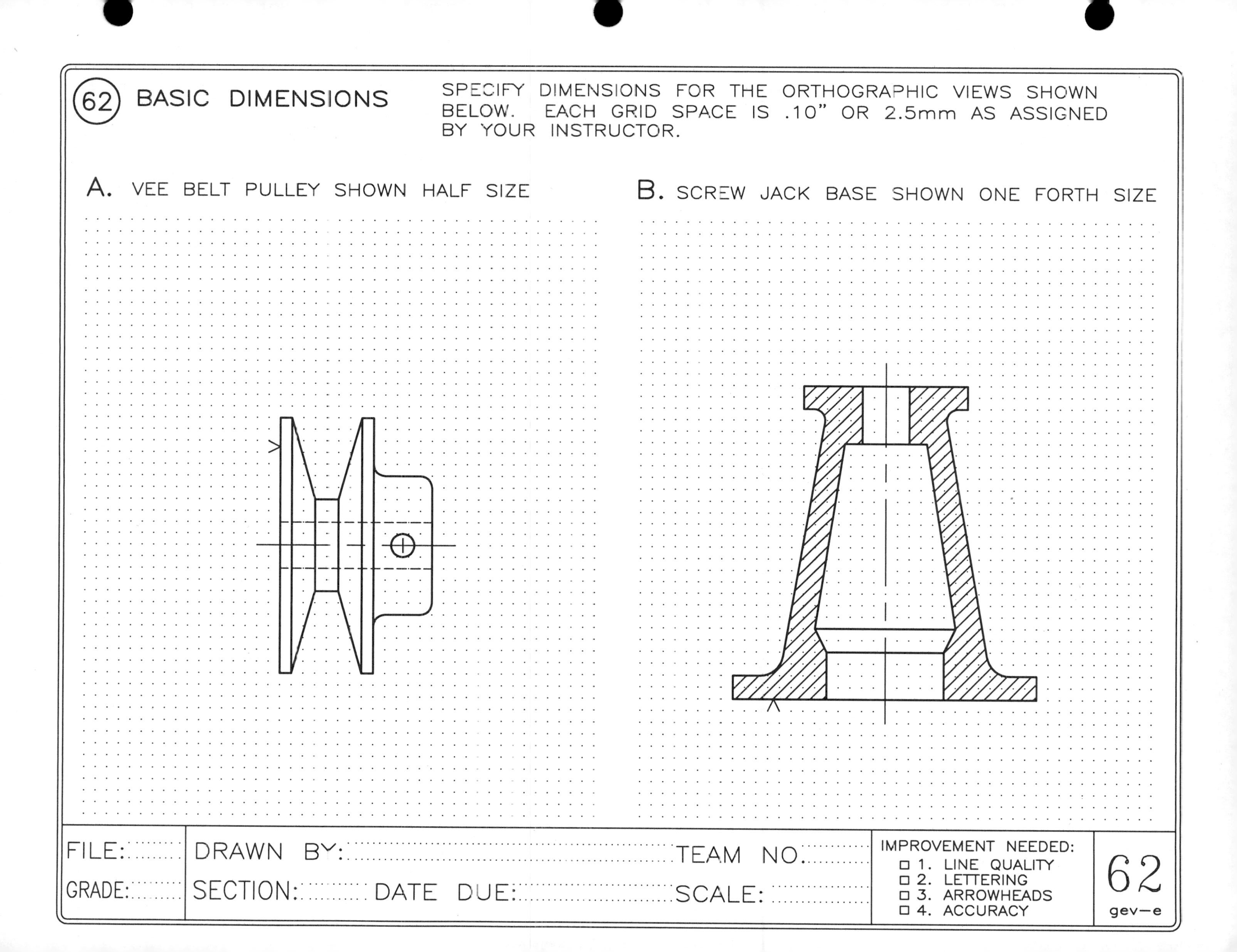
62
BASIC DIMENSIONS
SPECIFY DIMENSIONS FOR THE ORTHOGRAPHIC VIEWS SHOWN BELOW. EACH GRID SPACE IS .10" OR 2.5mm AS ASSIGNED BY YOUR INSTRUCTOR.
A. VEE BELT PULLEY SHOWN HALF SIZE
B. SCREW JACK BASE SHOWN ONE FORTH SIZE
FILE:
GRADE:
DRAWN BY:
SECTION:
DATE DUE:
TEAM NO.
SCALE:
IMPROVEMENT NEEDED:
1. LINE QUALITY
2. LETTERING
3. ARROWHEADS
4. ACCURACY
62
gev-e

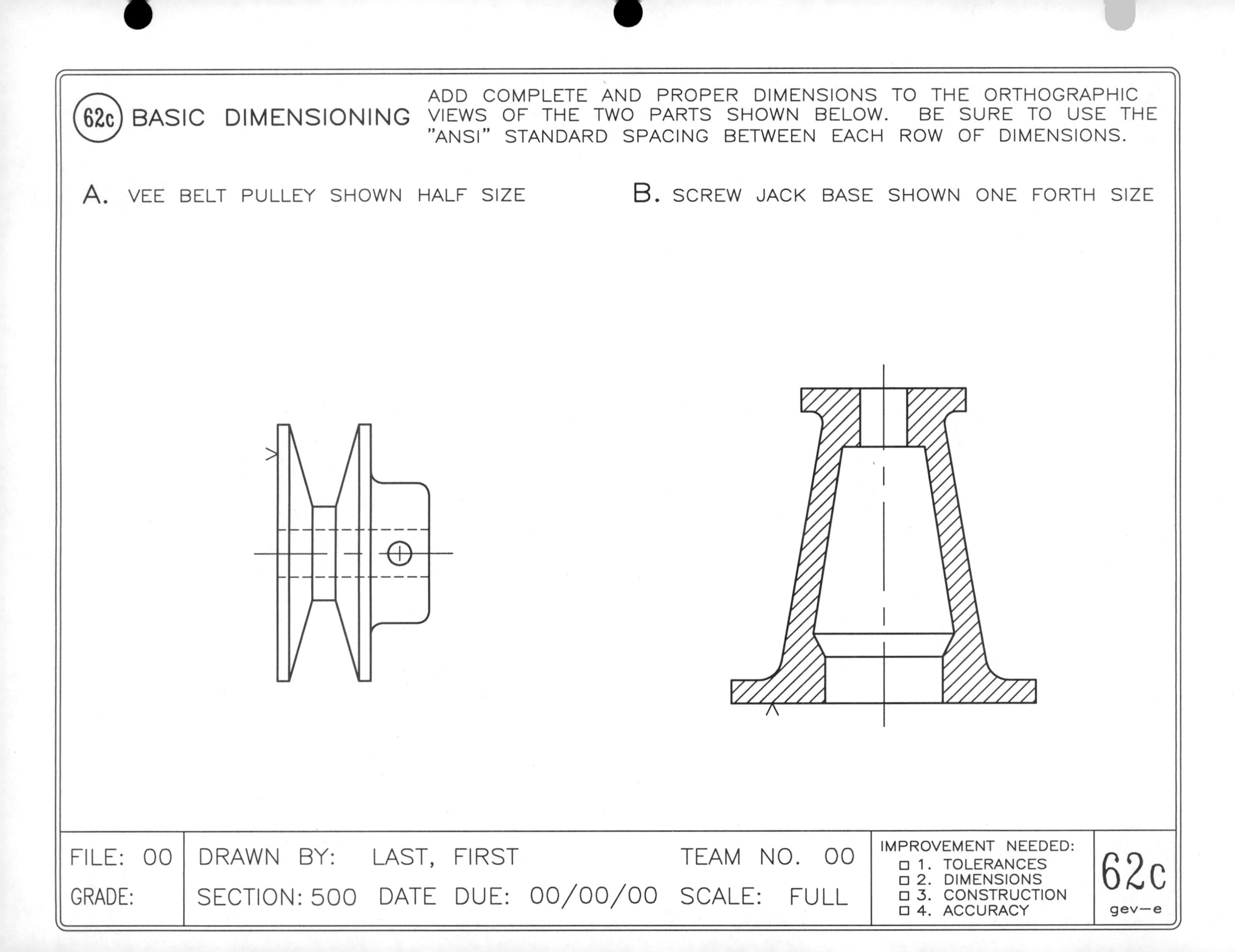
62c
BASIC DIMENSIONING
ADD COMPLETE AND PROPER DIMENSIONS TO THE ORTHOGRAPHIC VIEWS OF THE TWO PARTS SHOWN BELOW. BE SURE TO USE THE "ANSI" STANDARD SPACING BETWEEN EACH ROW OF DIMENSIONS.
A. VEE BELT PULLEY SHOWN HALF SIZE
B. SCREW JACK BASE SHOWN ONE FORTH SIZE
FILE: 00
GRADE:
DRAWN BY: LAST, FIRST
TEAM NO. 00
SECTION: 500
DATE DUE: 00/00/00
SCALE: FULL
IMPROVEMENT NEEDED:
☐ 1. TOLERANCES
☐ 2. DIMENSIONS
☐ 3. CONSTRUCTION
☐ 4. ACCURACY
62c
gev—e

63 BASIC DIMENSIONS

SPECIFY DIMENSIONS FOR THE ORTHOGRAPHIC VIEWS SHOWN BELOW. EACH GRID SPACE IS .10" OR 2.5mm AS ASSIGNED BY YOUR INSTRUCTOR.

A. VACUUM DRAIN PLUG WITH DIAMOND KNURL

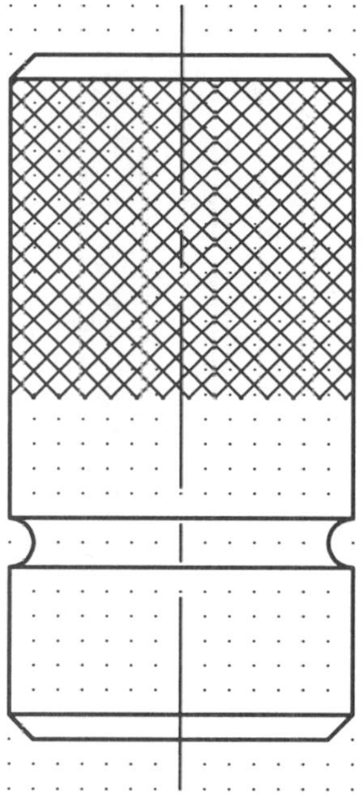

B. FLOATER LOCK TAB FOR CARBURETOR

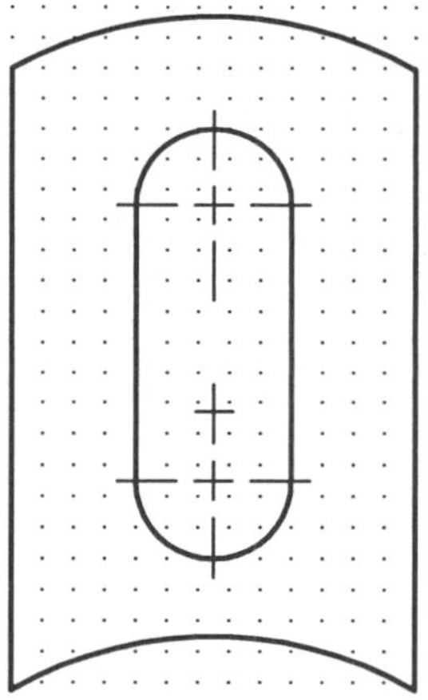

FILE:	DRAWN BY:	TEAM NO.	IMPROVEMENT NEEDED: ☐ 1. LINE QUALITY ☐ 2. LETTERING ☐ 3. ARROWHEADS ☐ 4. ACCURACY	63 gev–e
GRADE:	SECTION: DATE DUE:	SCALE:		

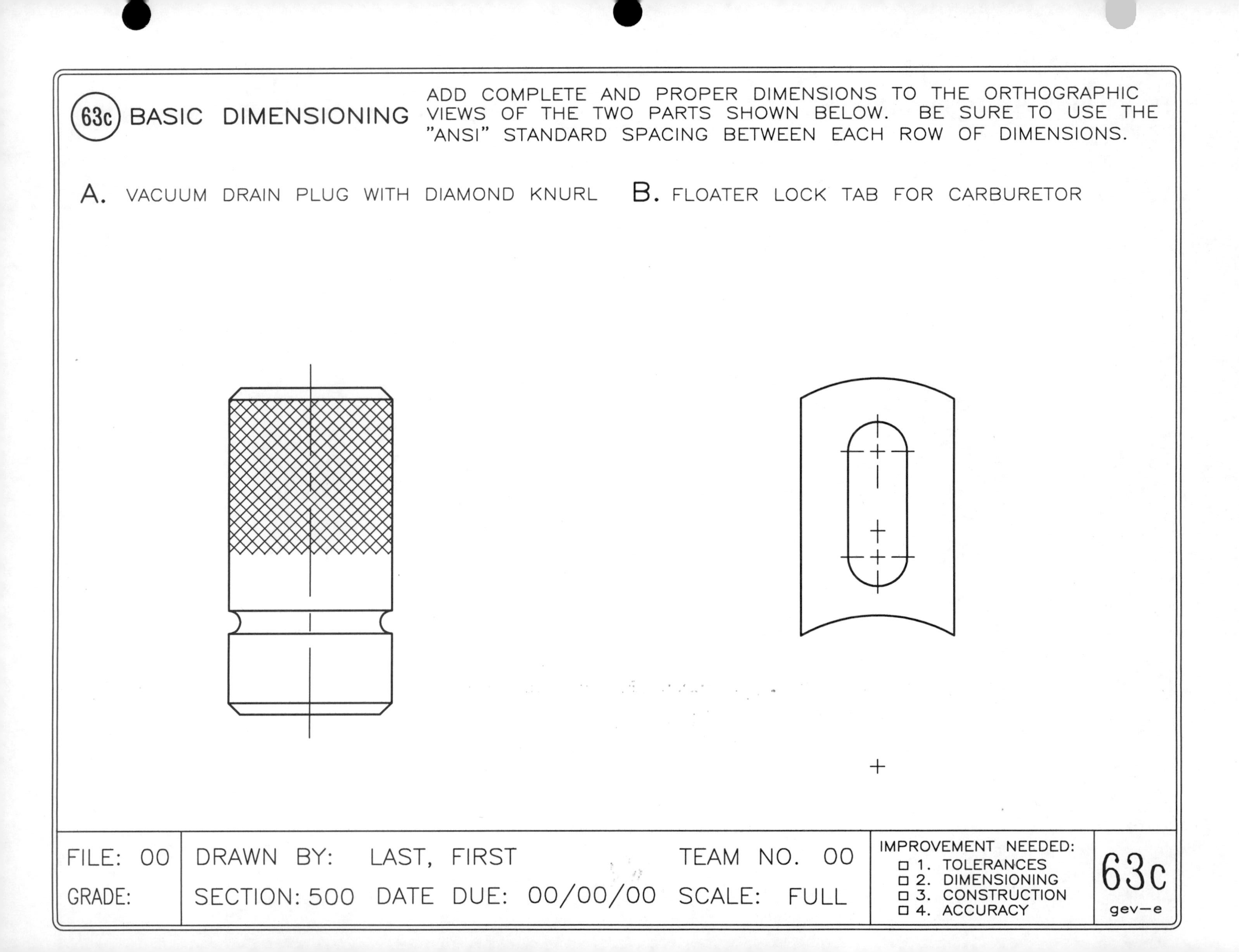
63c
BASIC DIMENSIONING
ADD COMPLETE AND PROPER DIMENSIONS TO THE ORTHOGRAPHIC VIEWS OF THE TWO PARTS SHOWN BELOW. BE SURE TO USE THE "ANSI" STANDARD SPACING BETWEEN EACH ROW OF DIMENSIONS.
A. VACUUM DRAIN PLUG WITH DIAMOND KNURL
B. FLOATER LOCK TAB FOR CARBURETOR
FILE: 00
GRADE:
DRAWN BY: LAST, FIRST
TEAM NO. 00
SECTION: 500
DATE DUE: 00/00/00
SCALE: FULL
IMPROVEMENT NEEDED:
□ 1. TOLERANCES
□ 2. DIMENSIONING
□ 3. CONSTRUCTION
□ 4. ACCURACY
63c
gev—e

(65) CYLINDRICAL FITS (mm)

REFER TO THE ANSI TABLE NOTED TO DIMENSION EACH METRIC HOLE AND CYLINDER. CALCULATE THE TOLERANCE, CLEARANCE AND ALLOWANCE OF EACH.

EXAMPLE:

	H11/c11 TABLE ø12 (NOMINAL)	H8/f7 ø20 (NOMINAL)	H7/g6 ø25 (NOMINAL)	H7/u6 ø30 (NOMINAL)
HOLE TOL:	.110			
SHFT. TOL:	.110			
ALLOWANCE:	.095			
CLEARANCE:	.315			

Ø11.905 / 11.795

Ø12.110 / 12.000

FILE: DRAWN BY: TEAM NO.
GRADE: SECTION: DATE DUE: SCALE:

IMPROVEMENT NEEDED:
- ☐ 1. LINE QUALITY
- ☐ 2. LETTERING
- ☐ 3. ARROWHEADS
- ☐ 4. ACCURACY

65
gev–d

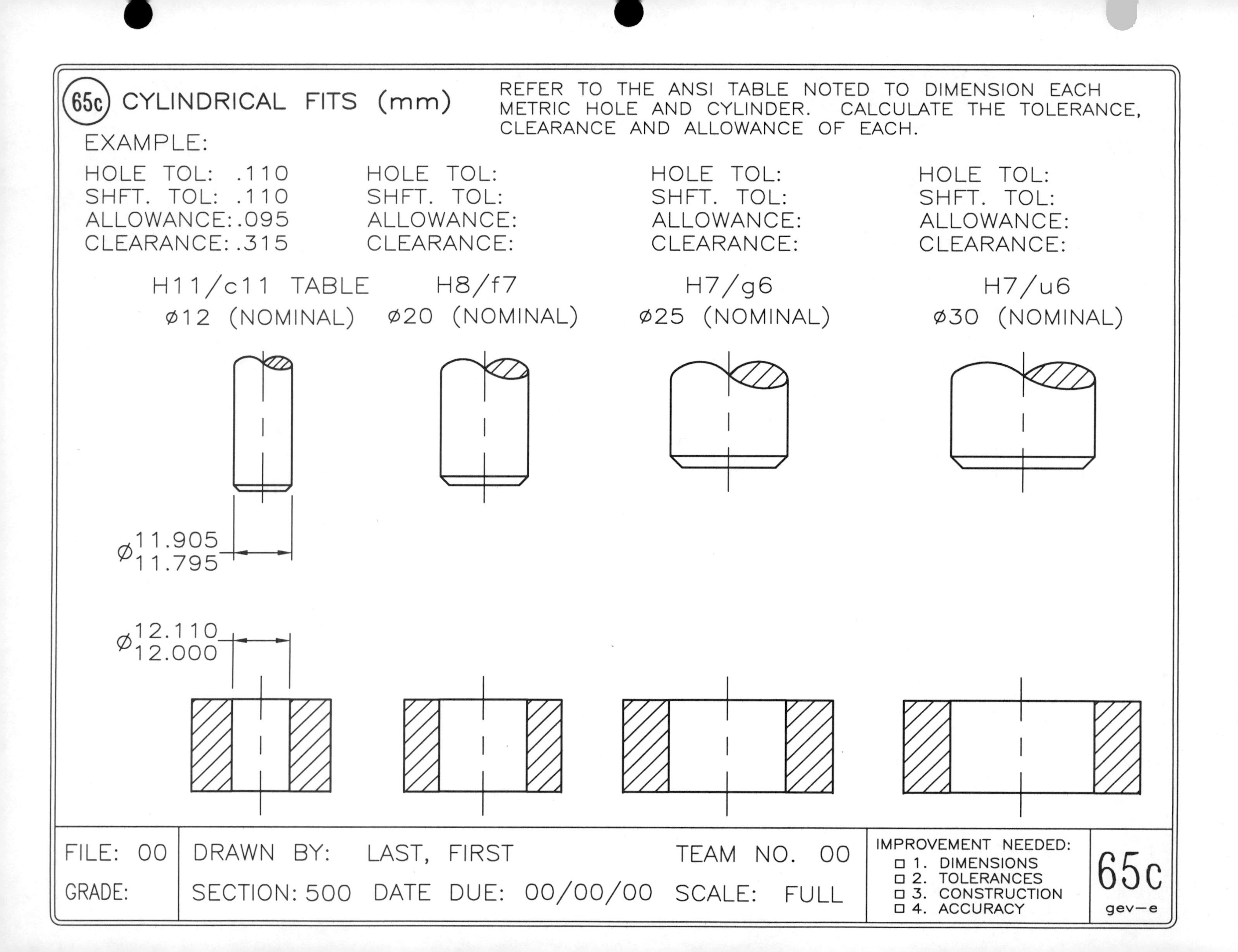
65c
CYLINDRICAL FITS (mm)
REFER TO THE ANSI TABLE NOTED TO DIMENSION EACH METRIC HOLE AND CYLINDER. CALCULATE THE TOLERANCE, CLEARANCE AND ALLOWANCE OF EACH.
EXAMPLE:
HOLE TOL: .110
SHFT. TOL: .110
ALLOWANCE: .095
CLEARANCE: .315
HOLE TOL:
SHFT. TOL:
ALLOWANCE:
CLEARANCE:
HOLE TOL:
SHFT. TOL:
ALLOWANCE:
CLEARANCE:
HOLE TOL:
SHFT. TOL:
ALLOWANCE:
CLEARANCE:
H11/c11 TABLE
ø12 (NOMINAL)
H8/f7
ø20 (NOMINAL)
H7/g6
ø25 (NOMINAL)
H7/u6
ø30 (NOMINAL)
Ø11.905
11.795
Ø12.110
12.000
FILE: 00
GRADE:
DRAWN BY: LAST, FIRST
SECTION: 500 DATE DUE: 00/00/00
TEAM NO. 00
SCALE: FULL
IMPROVEMENT NEEDED:
□ 1. DIMENSIONS
□ 2. TOLERANCES
□ 3. CONSTRUCTION
□ 4. ACCURACY
65c
gev-e

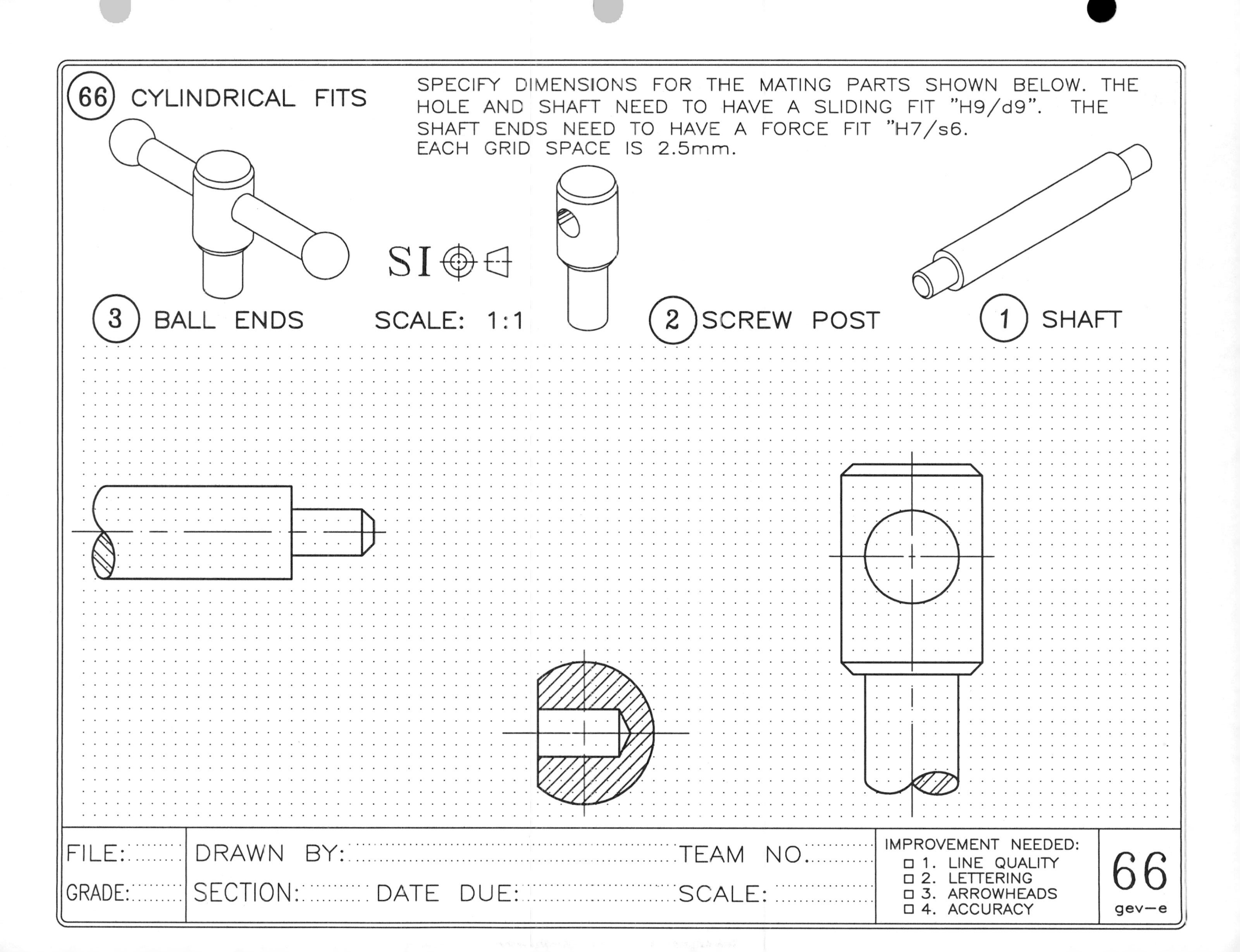
66 CYLINDRICAL FITS
SPECIFY DIMENSIONS FOR THE MATING PARTS SHOWN BELOW. THE HOLE AND SHAFT NEED TO HAVE A SLIDING FIT "H9/d9". THE SHAFT ENDS NEED TO HAVE A FORCE FIT "H7/s6.
EACH GRID SPACE IS 2.5mm.
SI
3 BALL ENDS
SCALE: 1:1
2 SCREW POST
1 SHAFT
FILE:
GRADE:
DRAWN BY:
TEAM NO.
SECTION:
DATE DUE:
SCALE:
IMPROVEMENT NEEDED:
1. LINE QUALITY
2. LETTERING
3. ARROWHEADS
4. ACCURACY
66
gev-e

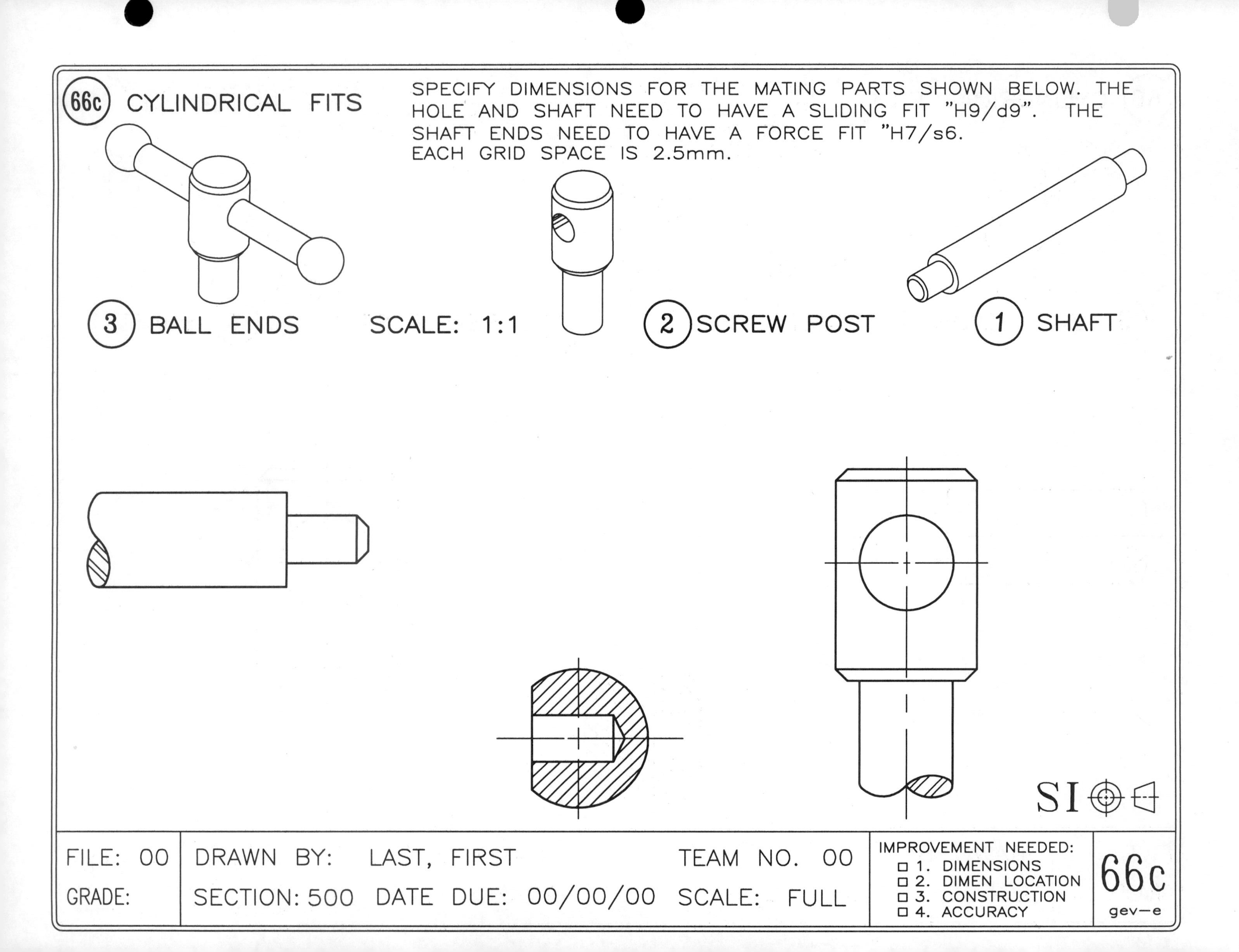
66c
CYLINDRICAL FITS
SPECIFY DIMENSIONS FOR THE MATING PARTS SHOWN BELOW. THE HOLE AND SHAFT NEED TO HAVE A SLIDING FIT "H9/d9". THE SHAFT ENDS NEED TO HAVE A FORCE FIT "H7/s6.
EACH GRID SPACE IS 2.5mm.
3 BALL ENDS
SCALE: 1:1
2 SCREW POST
1 SHAFT
SI
FILE: 00
GRADE:
DRAWN BY: LAST, FIRST
TEAM NO. 00
SECTION: 500
DATE DUE: 00/00/00
SCALE: FULL
IMPROVEMENT NEEDED:
☐ 1. DIMENSIONS
☐ 2. DIMEN LOCATION
☐ 3. CONSTRUCTION
☐ 4. ACCURACY
66c
gev—e

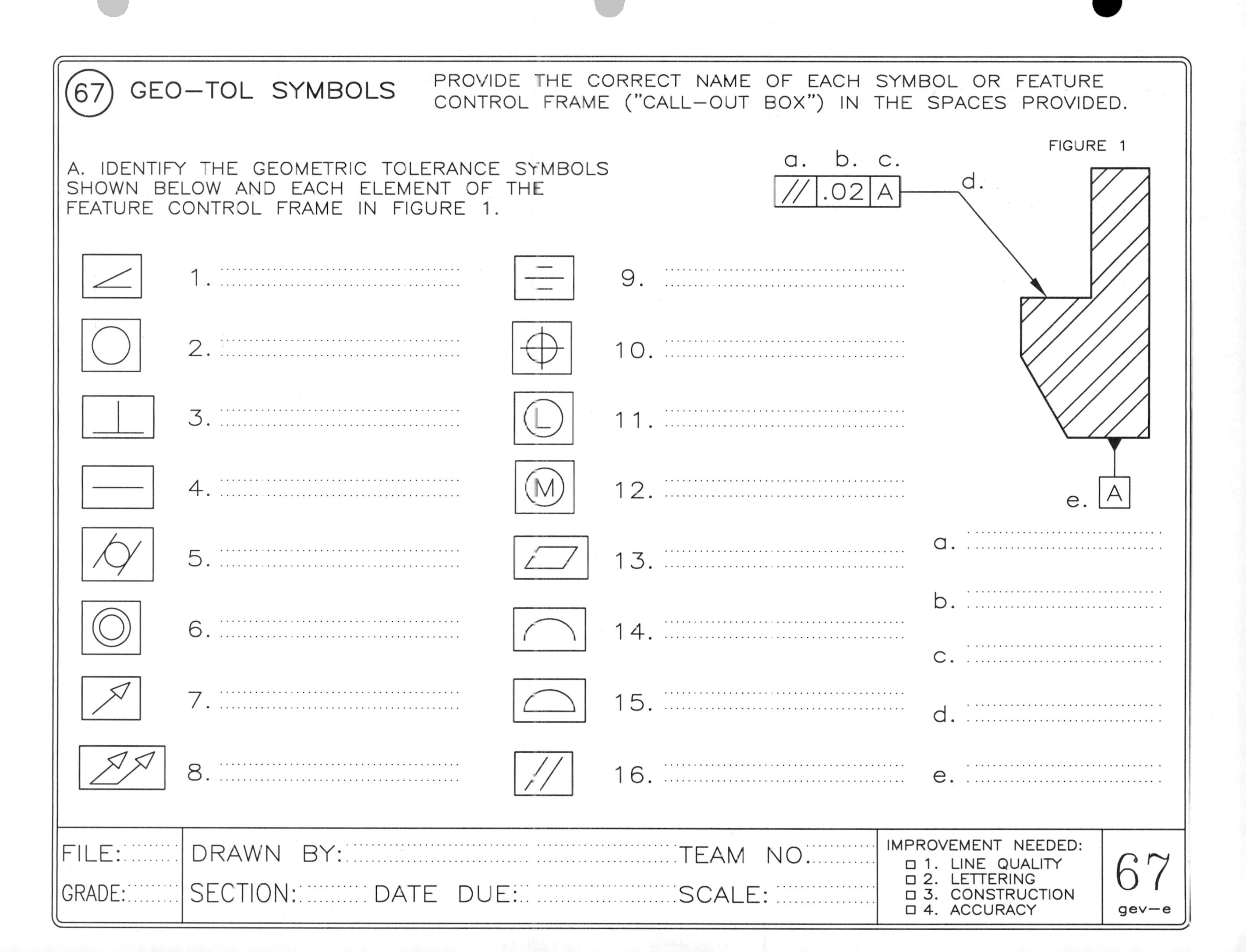

67
GEO—TOL SYMBOLS
PROVIDE THE CORRECT NAME OF EACH SYMBOL OR FEATURE CONTROL FRAME ("CALL—OUT BOX") IN THE SPACES PROVIDED.
A. IDENTIFY THE GEOMETRIC TOLERANCE SYMBOLS SHOWN BELOW AND EACH ELEMENT OF THE FEATURE CONTROL FRAME IN FIGURE 1.
FIGURE 1
a. b. c.
.02
A
d.
e.
A
1.
2.
3.
4.
5.
6.
7.
8.
L
M
9.
10.
11.
12.
13.
14.
15.
16.
a.
b.
c.
d.
e.
FILE:
GRADE:
DRAWN BY:
TEAM NO.
SECTION:
DATE DUE:
SCALE:
IMPROVEMENT NEEDED:
1. LINE QUALITY
2. LETTERING
3. CONSTRUCTION
4. ACCURACY
67
gev—e

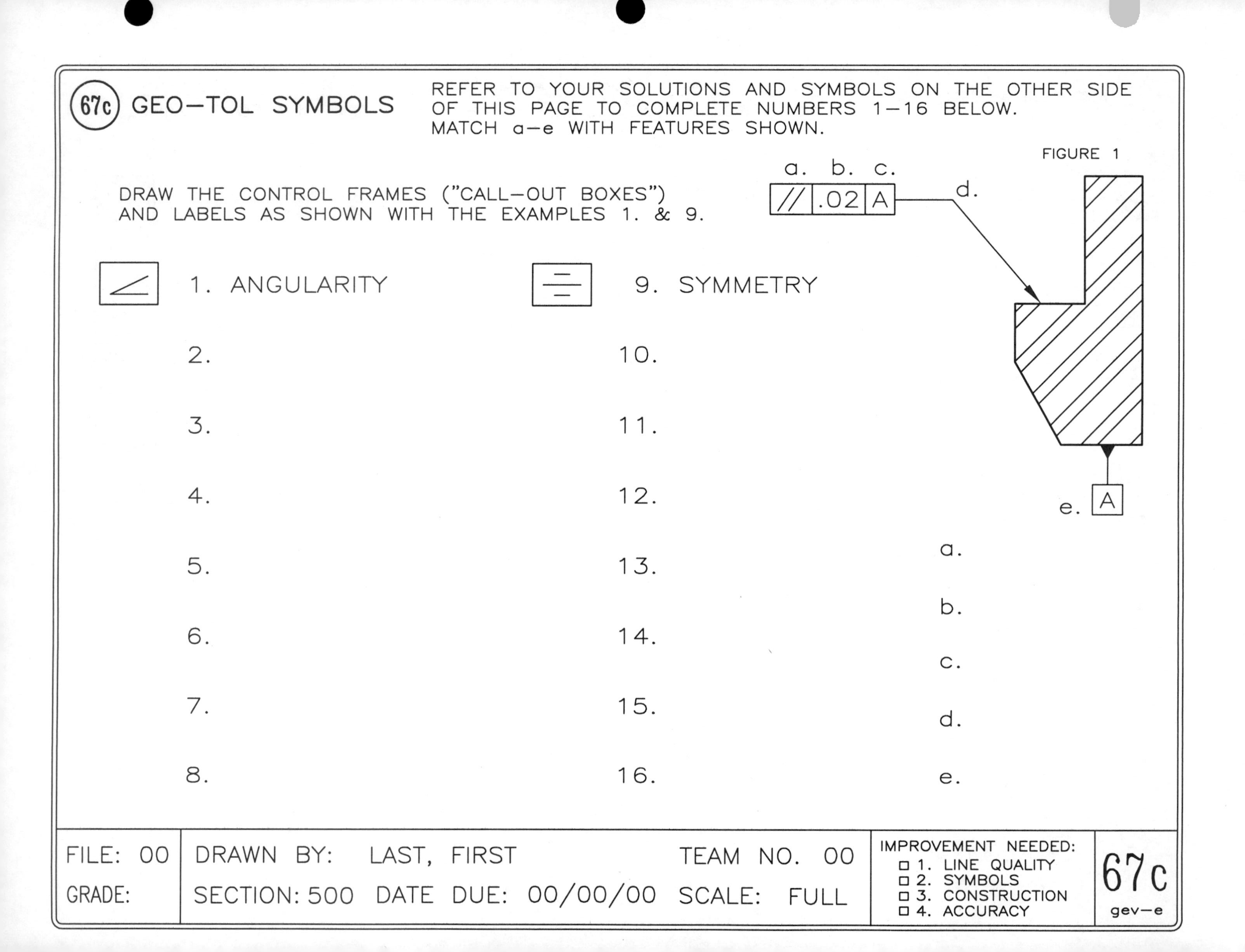
67c
GEO—TOL SYMBOLS
REFER TO YOUR SOLUTIONS AND SYMBOLS ON THE OTHER SIDE OF THIS PAGE TO COMPLETE NUMBERS 1—16 BELOW. MATCH a—e WITH FEATURES SHOWN.
FIGURE 1
a. b. c.
.02
A
d.
e.
A
DRAW THE CONTROL FRAMES ("CALL—OUT BOXES") AND LABELS AS SHOWN WITH THE EXAMPLES 1. & 9.
1. ANGULARITY
2.
3.
4.
5.
6.
7.
8.
9. SYMMETRY
10.
11.
12.
13.
14.
15.
16.
a.
b.
c.
d.
e.
FILE: 00
GRADE:
DRAWN BY: LAST, FIRST
TEAM NO. 00
SECTION: 500
DATE DUE: 00/00/00
SCALE: FULL
IMPROVEMENT NEEDED:
1. LINE QUALITY
2. SYMBOLS
3. CONSTRUCTION
4. ACCURACY
67c
gev—e

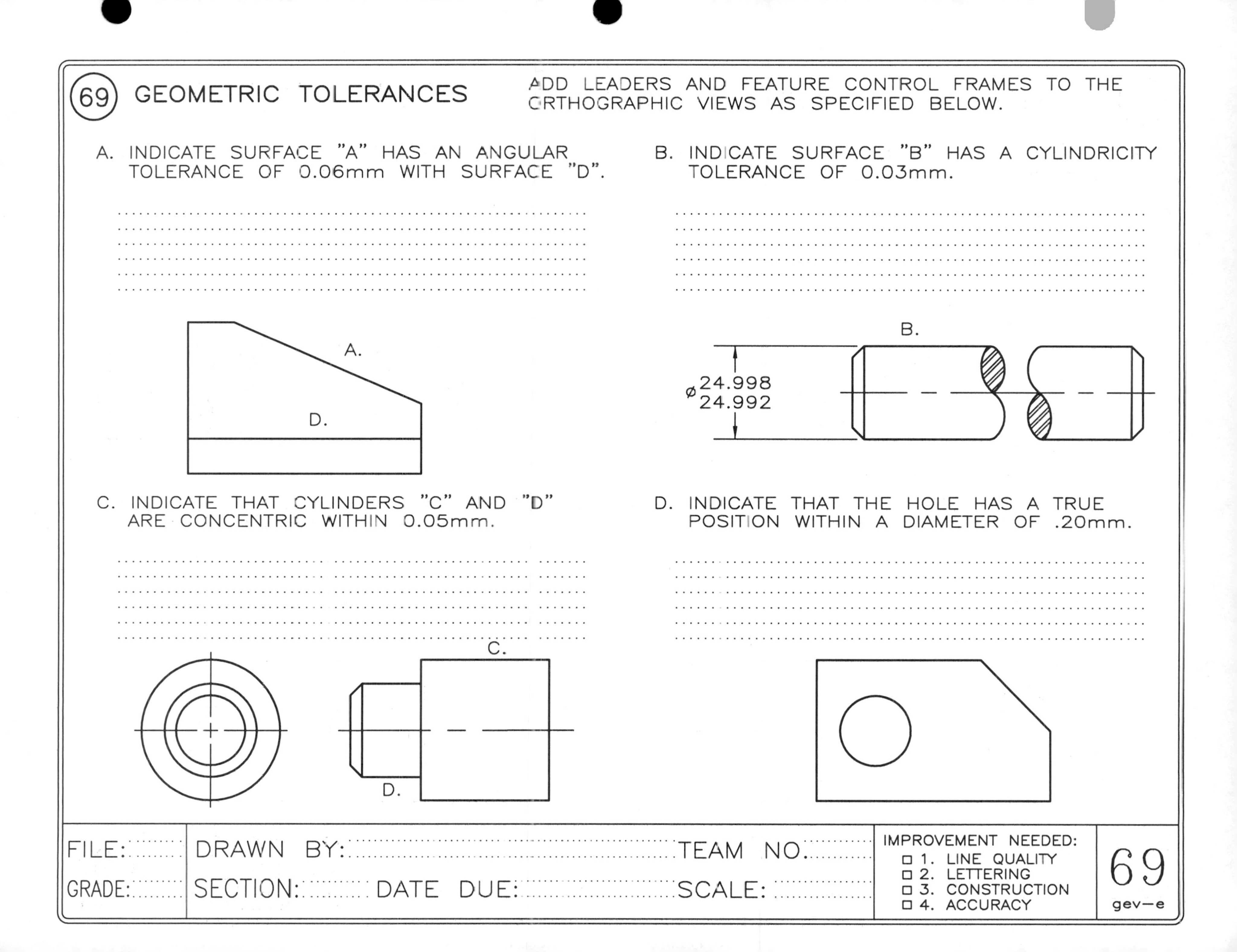
69
GEOMETRIC TOLERANCES
ADD LEADERS AND FEATURE CONTROL FRAMES TO THE ORTHOGRAPHIC VIEWS AS SPECIFIED BELOW.
A. INDICATE SURFACE "A" HAS AN ANGULAR TOLERANCE OF 0.06mm WITH SURFACE "D".
A.
D.
B. INDICATE SURFACE "B" HAS A CYLINDRICITY TOLERANCE OF 0.03mm.
B.
ø24.998
24.992
C. INDICATE THAT CYLINDERS "C" AND "D" ARE CONCENTRIC WITHIN 0.05mm.
C.
D.
D. INDICATE THAT THE HOLE HAS A TRUE POSITION WITHIN A DIAMETER OF .20mm.
FILE:
GRADE:
DRAWN BY:
SECTION:
DATE DUE:
TEAM NO.
SCALE:
IMPROVEMENT NEEDED:
1. LINE QUALITY
2. LETTERING
3. CONSTRUCTION
4. ACCURACY
69
gev-e

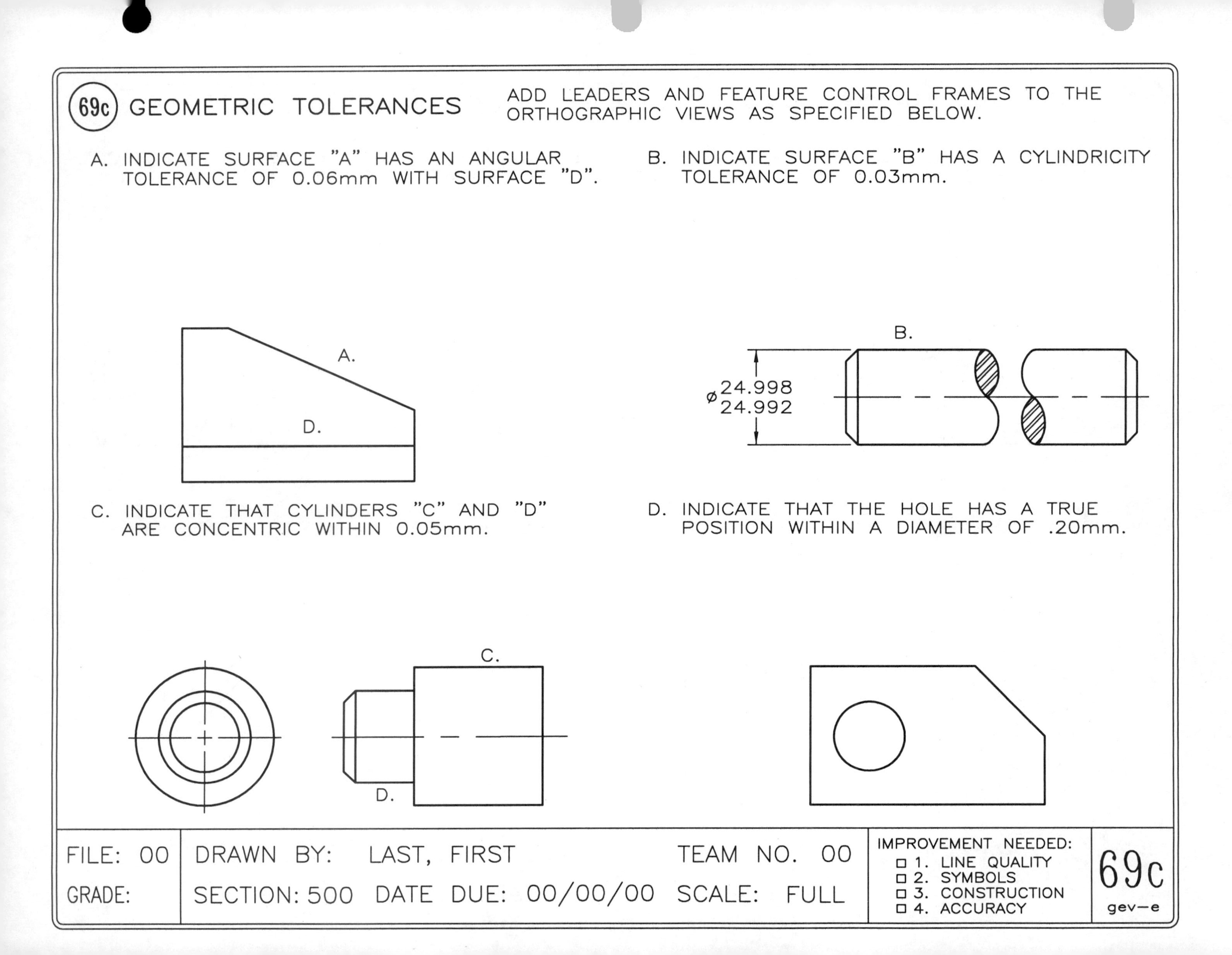
69c
GEOMETRIC TOLERANCES
ADD LEADERS AND FEATURE CONTROL FRAMES TO THE ORTHOGRAPHIC VIEWS AS SPECIFIED BELOW.
A. INDICATE SURFACE "A" HAS AN ANGULAR TOLERANCE OF 0.06mm WITH SURFACE "D".
B. INDICATE SURFACE "B" HAS A CYLINDRICITY TOLERANCE OF 0.03mm.
A.
D.
B.
ø24.998
24.992
C. INDICATE THAT CYLINDERS "C" AND "D" ARE CONCENTRIC WITHIN 0.05mm.
D. INDICATE THAT THE HOLE HAS A TRUE POSITION WITHIN A DIAMETER OF .20mm.
C.
D.
FILE: 00
GRADE:
DRAWN BY: LAST, FIRST
TEAM NO. 00
SECTION: 500
DATE DUE: 00/00/00
SCALE: FULL
IMPROVEMENT NEEDED:
□ 1. LINE QUALITY
□ 2. SYMBOLS
□ 3. CONSTRUCTION
□ 4. ACCURACY
69c
gev–e

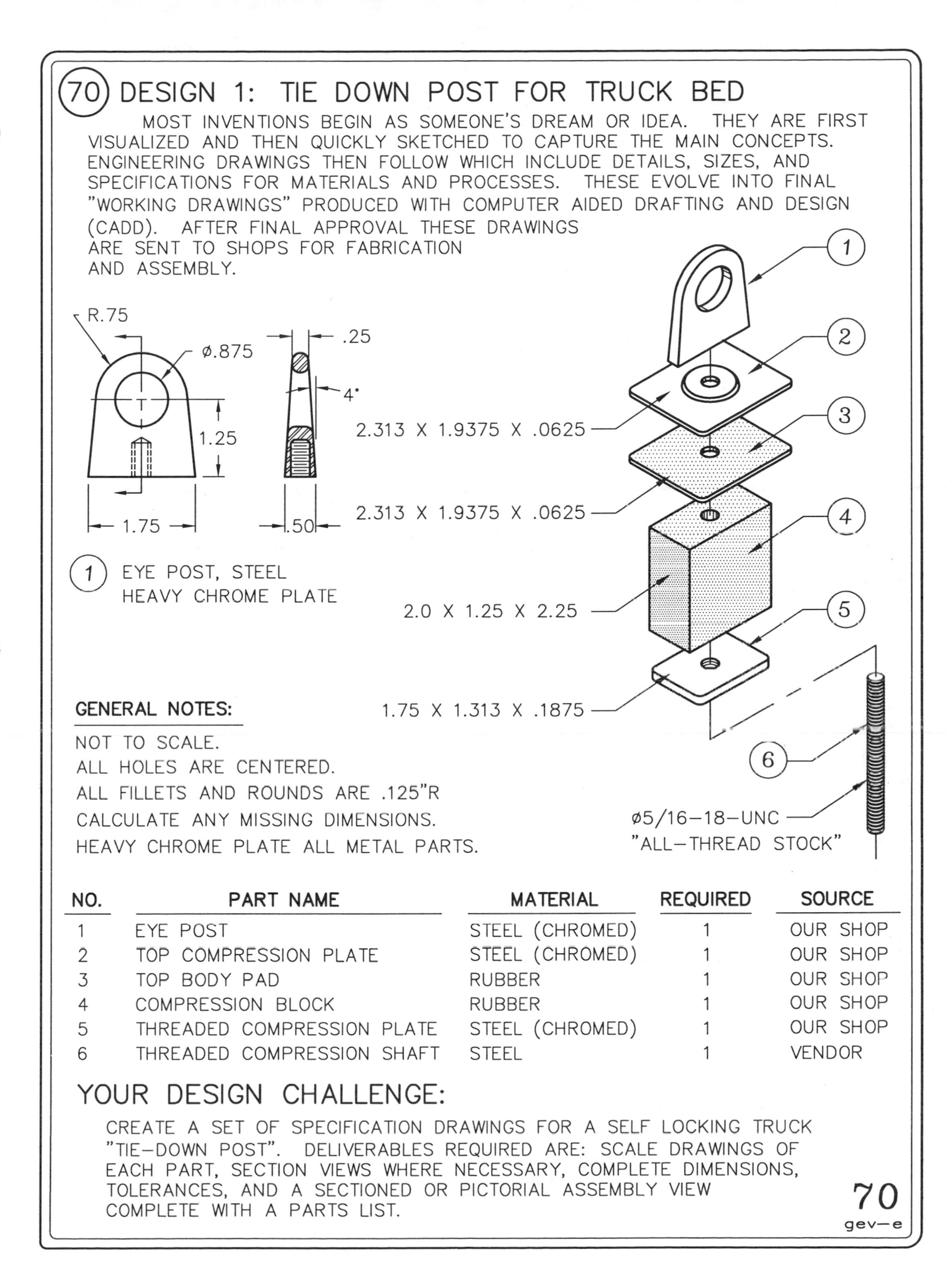

(70) DESIGN 1: TIE DOWN POST FOR TRUCK BED

MOST INVENTIONS BEGIN AS SOMEONE'S DREAM OR IDEA. THEY ARE FIRST VISUALIZED AND THEN QUICKLY SKETCHED TO CAPTURE THE MAIN CONCEPTS. ENGINEERING DRAWINGS THEN FOLLOW WHICH INCLUDE DETAILS, SIZES, AND SPECIFICATIONS FOR MATERIALS AND PROCESSES. THESE EVOLVE INTO FINAL "WORKING DRAWINGS" PRODUCED WITH COMPUTER AIDED DRAFTING AND DESIGN (CADD). AFTER FINAL APPROVAL THESE DRAWINGS ARE SENT TO SHOPS FOR FABRICATION AND ASSEMBLY.

(1) EYE POST, STEEL
HEAVY CHROME PLATE

GENERAL NOTES:

NOT TO SCALE.
ALL HOLES ARE CENTERED.
ALL FILLETS AND ROUNDS ARE .125"R
CALCULATE ANY MISSING DIMENSIONS.
HEAVY CHROME PLATE ALL METAL PARTS.

NO.	PART NAME	MATERIAL	REQUIRED	SOURCE
1	EYE POST	STEEL (CHROMED)	1	OUR SHOP
2	TOP COMPRESSION PLATE	STEEL (CHROMED)	1	OUR SHOP
3	TOP BODY PAD	RUBBER	1	OUR SHOP
4	COMPRESSION BLOCK	RUBBER	1	OUR SHOP
5	THREADED COMPRESSION PLATE	STEEL (CHROMED)	1	OUR SHOP
6	THREADED COMPRESSION SHAFT	STEEL	1	VENDOR

YOUR DESIGN CHALLENGE:

CREATE A SET OF SPECIFICATION DRAWINGS FOR A SELF LOCKING TRUCK "TIE–DOWN POST". DELIVERABLES REQUIRED ARE: SCALE DRAWINGS OF EACH PART, SECTION VIEWS WHERE NECESSARY, COMPLETE DIMENSIONS, TOLERANCES, AND A SECTIONED OR PICTORIAL ASSEMBLY VIEW COMPLETE WITH A PARTS LIST.

0 1 2 3 4 5 6 7 8 9

10 SCALE FULL SIZE INCHES

0 1 2 3 4 5 6 7 8 9 10 11 12 13 14 15 16 17 18

20 SCALE 1 INCH EQUALS 20 UNITS.

0 2 4 6 8 10 12 14 16 18 20 22 24 26

30 SCALE 1 INCH EQUALS 30 UNITS.

0 2 4 6 8 10 12 14 16 18 20 22 24 26 28 30 32 34 36

40 SCALE 1 INCH EQUALS 40 UNITS.

0 2 4 6 8 10 12 14 16 18 20 22 24 26 28 30 32 34 36 38 40 42 44

50 SCALE 1 INCH EQUALS 50 UNITS.

0 2 4 6 8 10 12 14 16 18 20 22 24 26 28 30 32 34 36 38 40 42 44 46 48 50 52 54

60 SCALE 1 INCH EQUALS 60 UNITS.

COMMON ENGINEERS' SCALES

USE THESE SCALES TO ESTABLISH DIMENSIONS ON DRAWING EXERCISES.

70c

gev–e

(71) DESIGN 2: TONGUE ASEMBLY FOR TRAILER HITCH

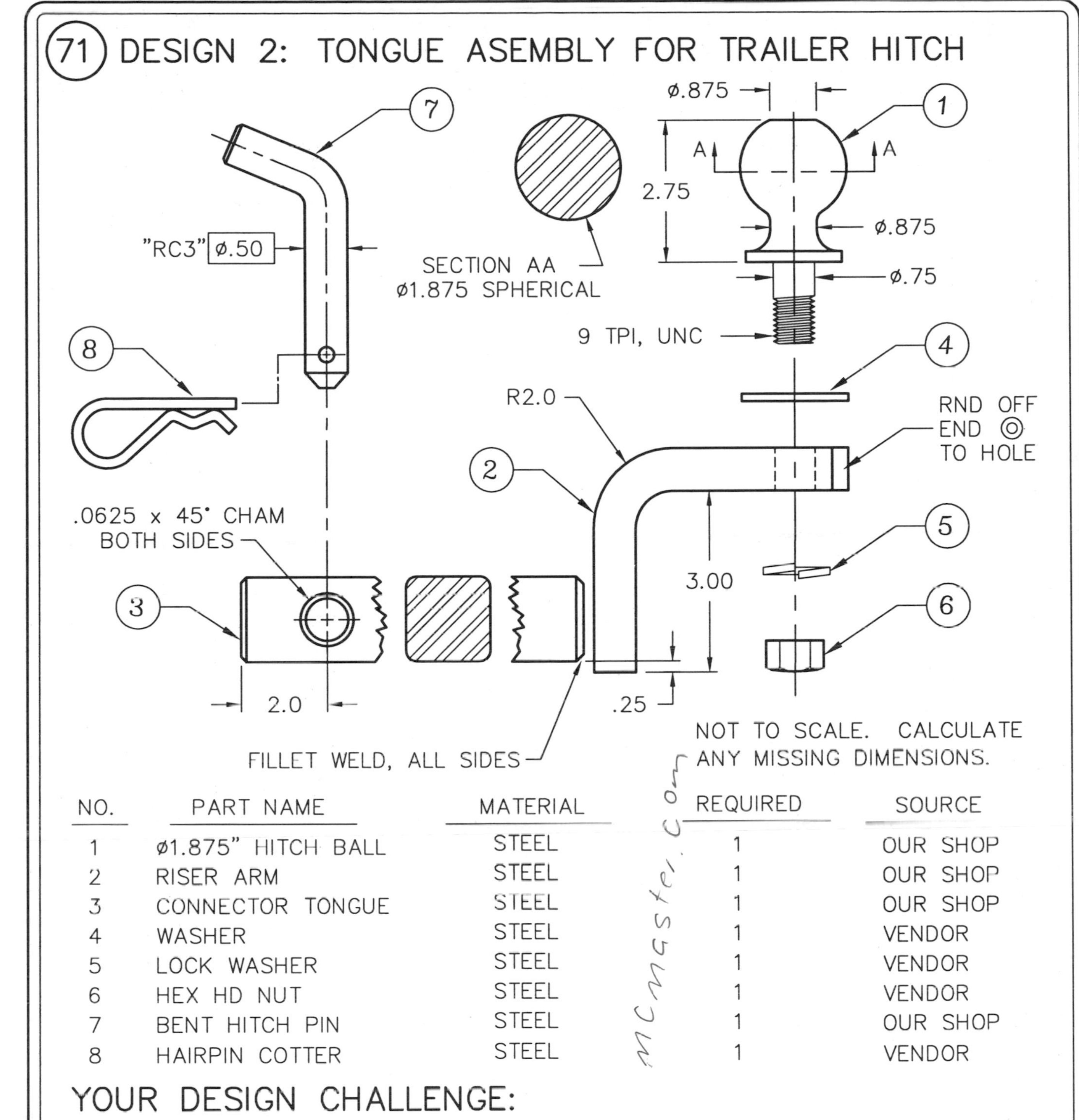

NOT TO SCALE. CALCULATE ANY MISSING DIMENSIONS.

NO.	PART NAME	MATERIAL	REQUIRED	SOURCE
1	ø1.875" HITCH BALL	STEEL	1	OUR SHOP
2	RISER ARM	STEEL	1	OUR SHOP
3	CONNECTOR TONGUE	STEEL	1	OUR SHOP
4	WASHER	STEEL	1	VENDOR
5	LOCK WASHER	STEEL	1	VENDOR
6	HEX HD NUT	STEEL	1	VENDOR
7	BENT HITCH PIN	STEEL	1	OUR SHOP
8	HAIRPIN COTTER	STEEL	1	VENDOR

YOUR DESIGN CHALLENGE:

CREATE A SET OF SPECIFICATION DRAWINGS FOR A "TRAILER HITCH" TONGUE ASSEMBLY. THE TONGUE IS A SOLID STEEL BAR THAT IS 1.25" SQ. AND 6.50" LONG. THE HOLE IN THE TONGUE IS A BASIC SIZE OF .50" WITH AN "RC3" FIT. THE TONGUE SHOULD BE TOLERANCED TO FIT INTO THE BUMPER RECEIVER AS A "RC2" FIT. THE RISER ARM IS MADE FROM A .625" THICK X 2.0" WIDE STEEL BAR. THE END OF THE RISER ARM CLOSEST TO THE BALL IS ROUNDED OFF TO BE CONCENTRIC TO THE HOLE. ALL PARTS HAVE A HEAVY CHROME FINISH.

DELIVERABLES REQUIRED ARE: (1) FULLY DIMENSIONED, SCALE DRAWINGS OF EACH PART WITH SECTION VIEWS AND TOLERANCES WHERE NECESSARY. (2) A SECTIONED OR PICTORIAL ASSEMBLY VIEW COMPLETE WITH A PARTS LIST. WORK ON A "B" SIZE SHEET OR ON THE WD SHEETS FOUND IN THIS BOOK.

0 1 2 3 4 5 6 7 8 9 10 11 12 13 14 15 16 17 18 19 20 21 22 23

1:1 SCALE FULL SIZE MILLIMETERS

0 5 10 15 20 25 30 35 40 45

1:2 SCALE METRIC SCALE

0 10 20 30 40 50 60

1:3 SCALE METRIC SCALE

0 10 20 30 40 50 60 70 80 90

1:4 SCALE METRIC SCALE

0 10 20 30 40 50 60 70 80 90 100 110

1:5 SCALE METRIC SCALE

0 10 20 30 40 50 60 70 80 90 100 120 130 130

1:6 SCALE METRIC SCALE

COMMON METRIC SCALES

USE THESE SCALES TO ESTABLISH DIMENSIONS ON DRAWING EXERCISES.

DESIGN 3: HEAVY DUTY SCREW JACK

SHOWN ONE-FOURTH SCALE. ESTABLISH ANY MISSING DIMENSIONS.
SEE DIMENSIONS AND DETAILS ON BACK.

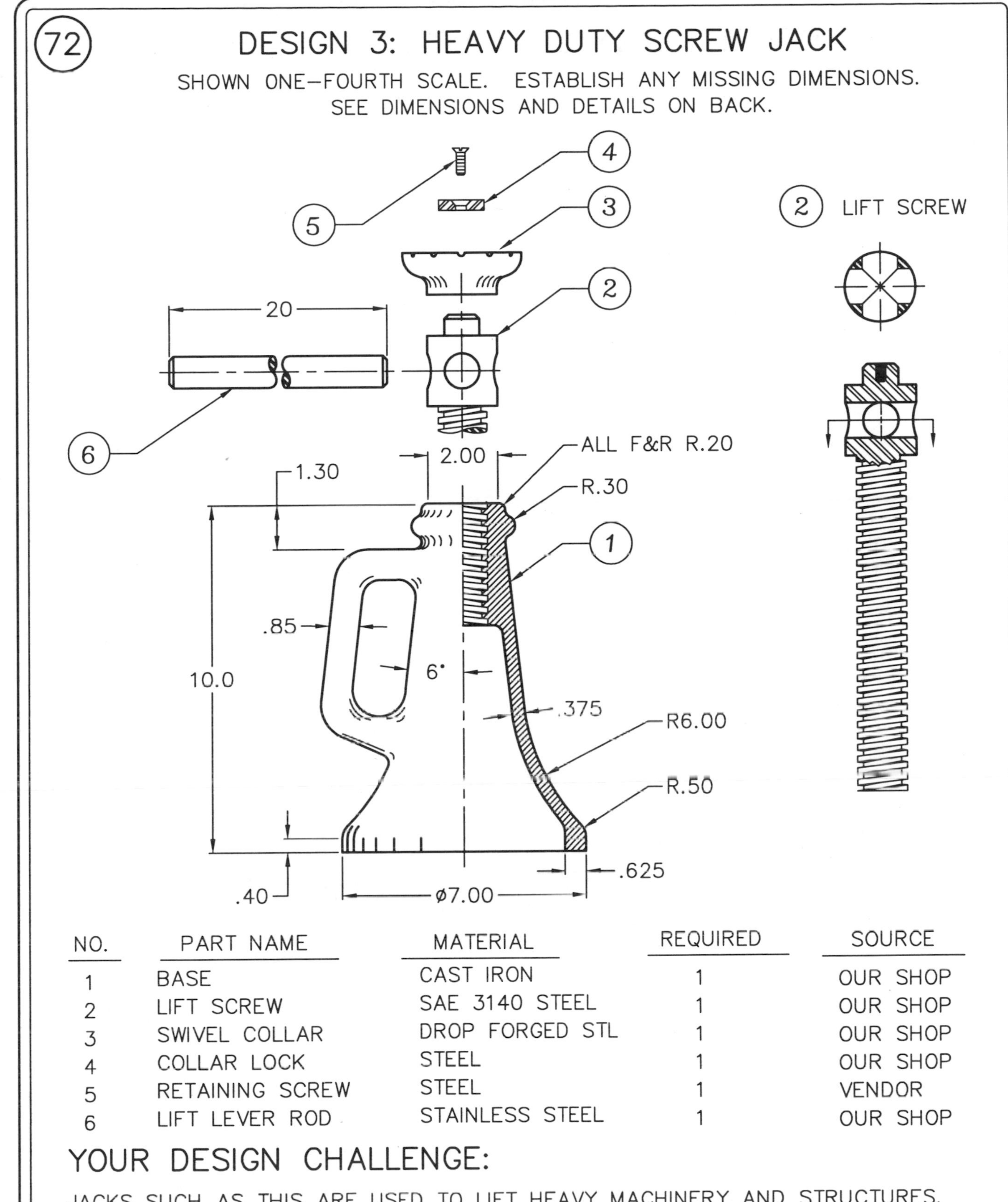

NO.	PART NAME	MATERIAL	REQUIRED	SOURCE
1	BASE	CAST IRON	1	OUR SHOP
2	LIFT SCREW	SAE 3140 STEEL	1	OUR SHOP
3	SWIVEL COLLAR	DROP FORGED STL	1	OUR SHOP
4	COLLAR LOCK	STEEL	1	OUR SHOP
5	RETAINING SCREW	STEEL	1	VENDOR
6	LIFT LEVER ROD	STAINLESS STEEL	1	OUR SHOP

YOUR DESIGN CHALLENGE:

JACKS SUCH AS THIS ARE USED TO LIFT HEAVY MACHINERY AND STRUCTURES. REDESIGN THE ONE SHOWN ABOVE TO ENABLE IT TO FIT UNDER MACHINERY THAT HAS ONLY A 12" CLEARANCE FROM THE SHOP FLOOR. DELIVER A SET OF DESIGN DRAWINGS THAT INCLUDE ALL DIMENSIONS REQUIRED FOR PRODUCTION AND AN ASSEMBLY VIEW COMPLETE WITH A PARTS LIST. LIST THE MAXIMUM LIFT HEIGHT OF YOUR DESIGN WITH SIX THREADS FULLY ENGAGED.

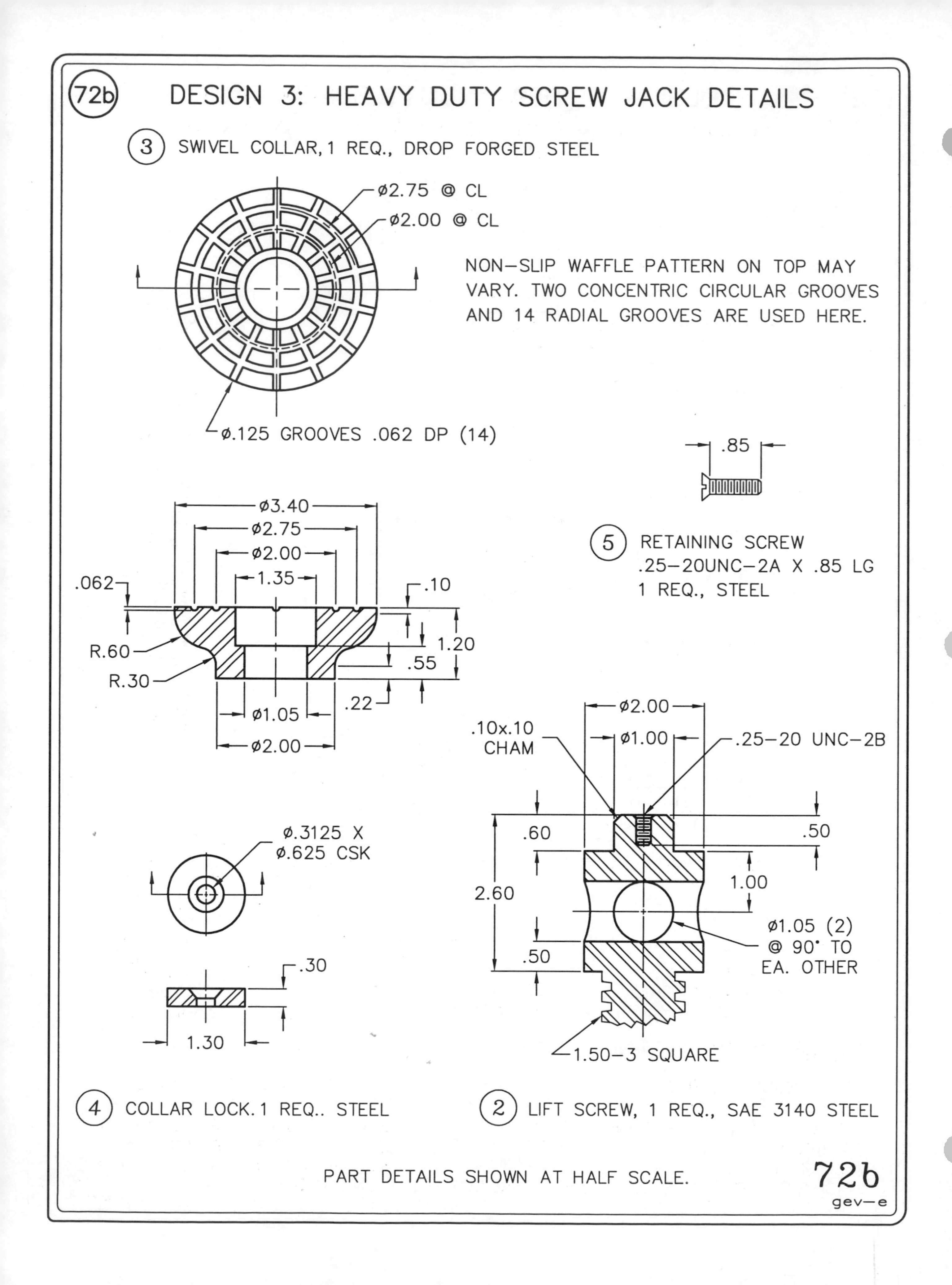

72b
DESIGN 3: HEAVY DUTY SCREW JACK DETAILS
3 SWIVEL COLLAR, 1 REQ., DROP FORGED STEEL
ø2.75 @ CL
ø2.00 @ CL
NON-SLIP WAFFLE PATTERN ON TOP MAY VARY. TWO CONCENTRIC CIRCULAR GROOVES AND 14 RADIAL GROOVES ARE USED HERE.
ø.125 GROOVES .062 DP (14)
.85
ø3.40
ø2.75
ø2.00
1.35
.062
.10
R.60
R.30
1.20
.55
.22
ø1.05
ø2.00
5 RETAINING SCREW
.25-20UNC-2A X .85 LG
1 REQ., STEEL
ø2.00
.10x.10
CHAM
ø1.00
.25-20 UNC-2B
.60
.50
1.00
2.60
ø1.05 (2)
@ 90° TO
EA. OTHER
.50
1.50-3 SQUARE
ø.3125 X
ø.625 CSK
.30
1.30
4 COLLAR LOCK. 1 REQ.. STEEL
2 LIFT SCREW, 1 REQ., SAE 3140 STEEL
PART DETAILS SHOWN AT HALF SCALE.
72b
gev-e

(74) AUXILIARY VIEWS

CONSTRUCT A TRUE SIZE AUXILIARY VIEW OF THE INCLINED EDGE OF THE DRILL JIG. INDICATE TRUE SIZE WITH "TS" AND LABEL ALL LINES AND PLANES.

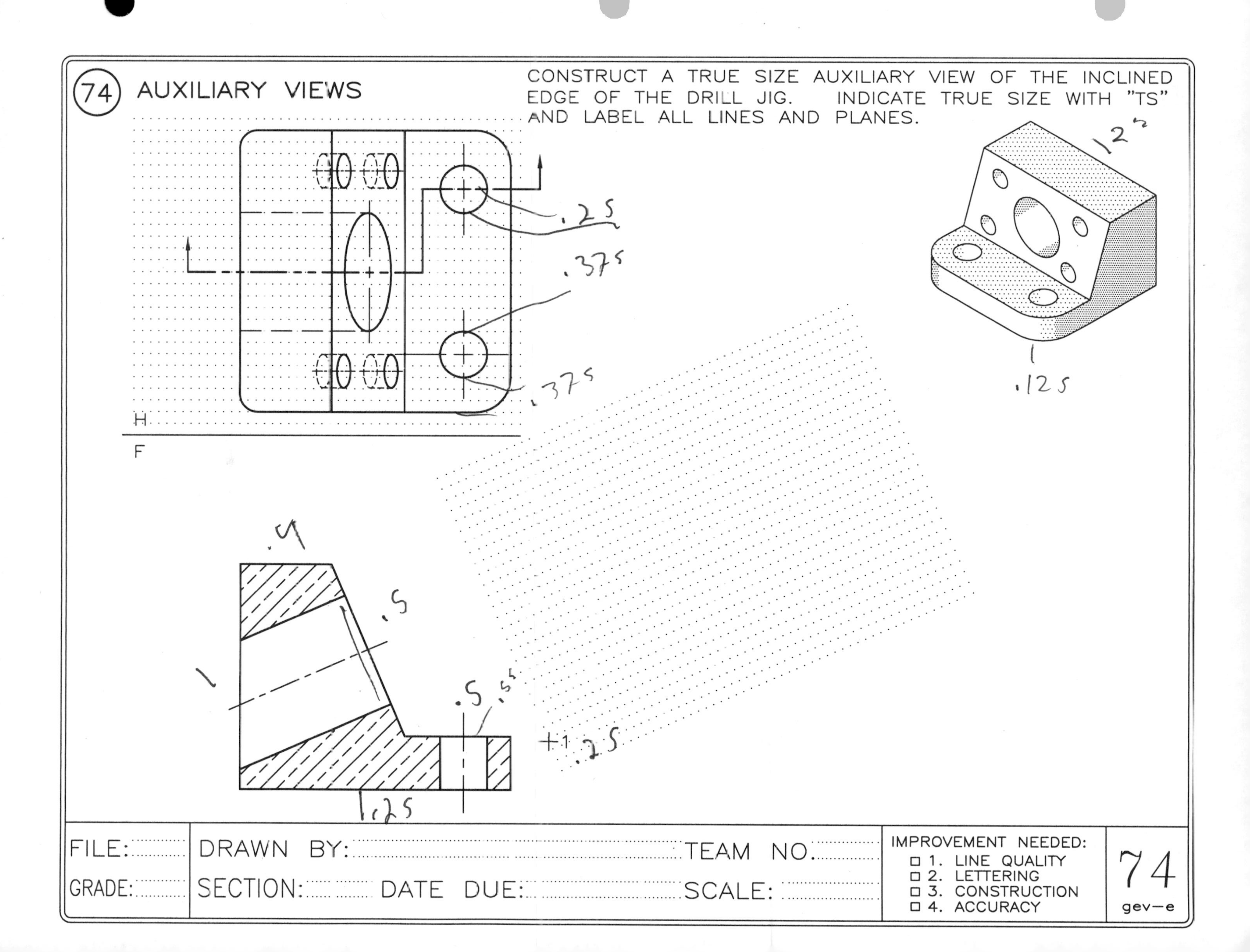

FILE: DRAWN BY: TEAM NO.

GRADE: SECTION: DATE DUE: SCALE:

IMPROVEMENT NEEDED:
- ☐ 1. LINE QUALITY
- ☐ 2. LETTERING
- ☐ 3. CONSTRUCTION
- ☐ 4. ACCURACY

74

gev–e

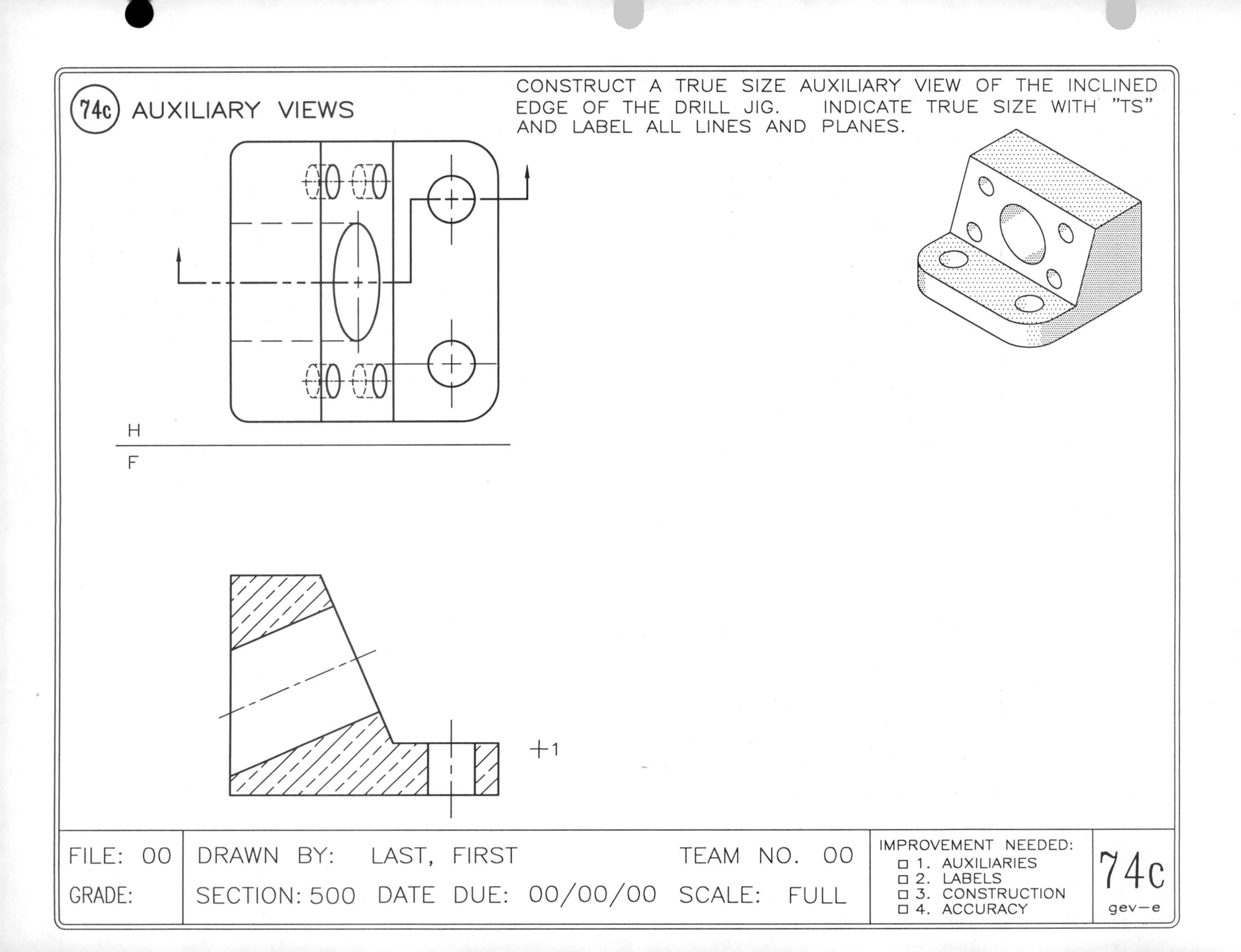
74c
AUXILIARY VIEWS
CONSTRUCT A TRUE SIZE AUXILIARY VIEW OF THE INCLINED EDGE OF THE DRILL JIG. INDICATE TRUE SIZE WITH "TS" AND LABEL ALL LINES AND PLANES.
H
F
1
FILE: 00
GRADE:
DRAWN BY: LAST, FIRST
TEAM NO. 00
SECTION: 500
DATE DUE: 00/00/00
SCALE: FULL
IMPROVEMENT NEEDED:
1. AUXILIARIES
2. LABELS
3. CONSTRUCTION
4. ACCURACY
74c
gev-e

(75) AUXILIARY VIEWS

CREATE A TRUE SIZE AUXILIARY VIEW OF THE HANDLEBAR MOUNTING BASE SHOWN BELOW. INDICATE THE TRUE SIZE PLANE WITH "TS" AND LABEL ALL REFERENCE LINES AND PLANES.

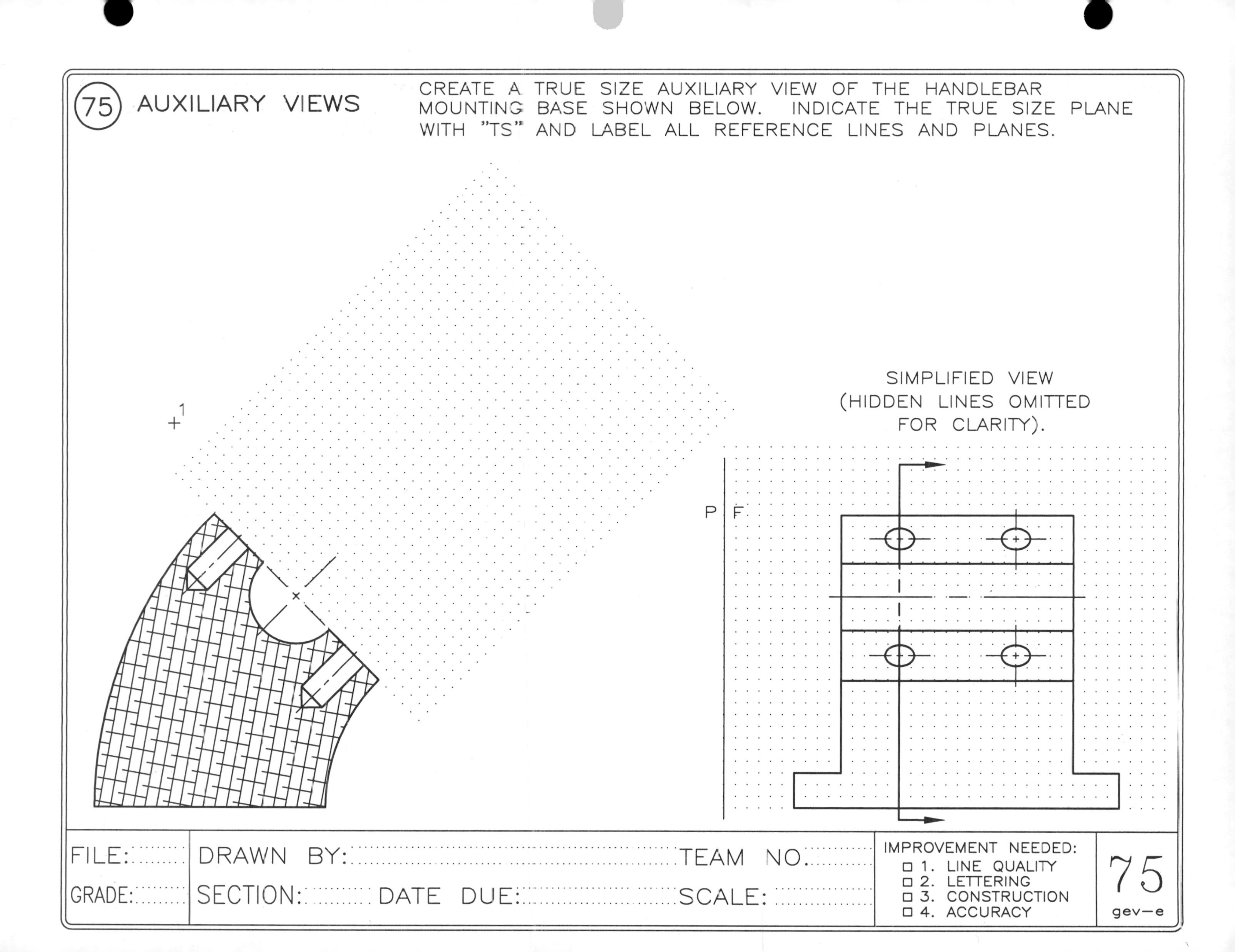

FILE:	DRAWN BY:	TEAM NO.	IMPROVEMENT NEEDED:	75
GRADE:	SECTION: DATE DUE:	SCALE:	□ 1. LINE QUALITY □ 2. LETTERING □ 3. CONSTRUCTION □ 4. ACCURACY	gev-e

75c AUXILIARY VIEWS

CREATE A TRUE SIZE AUXILIARY VIEW OF THE HANDLEBAR MOUNTING BASE SHOWN BELOW. INDICATE THE TRUE SIZE PLANE WITH "TS" AND LABEL ALL REFERENCE LINES AND PLANES.

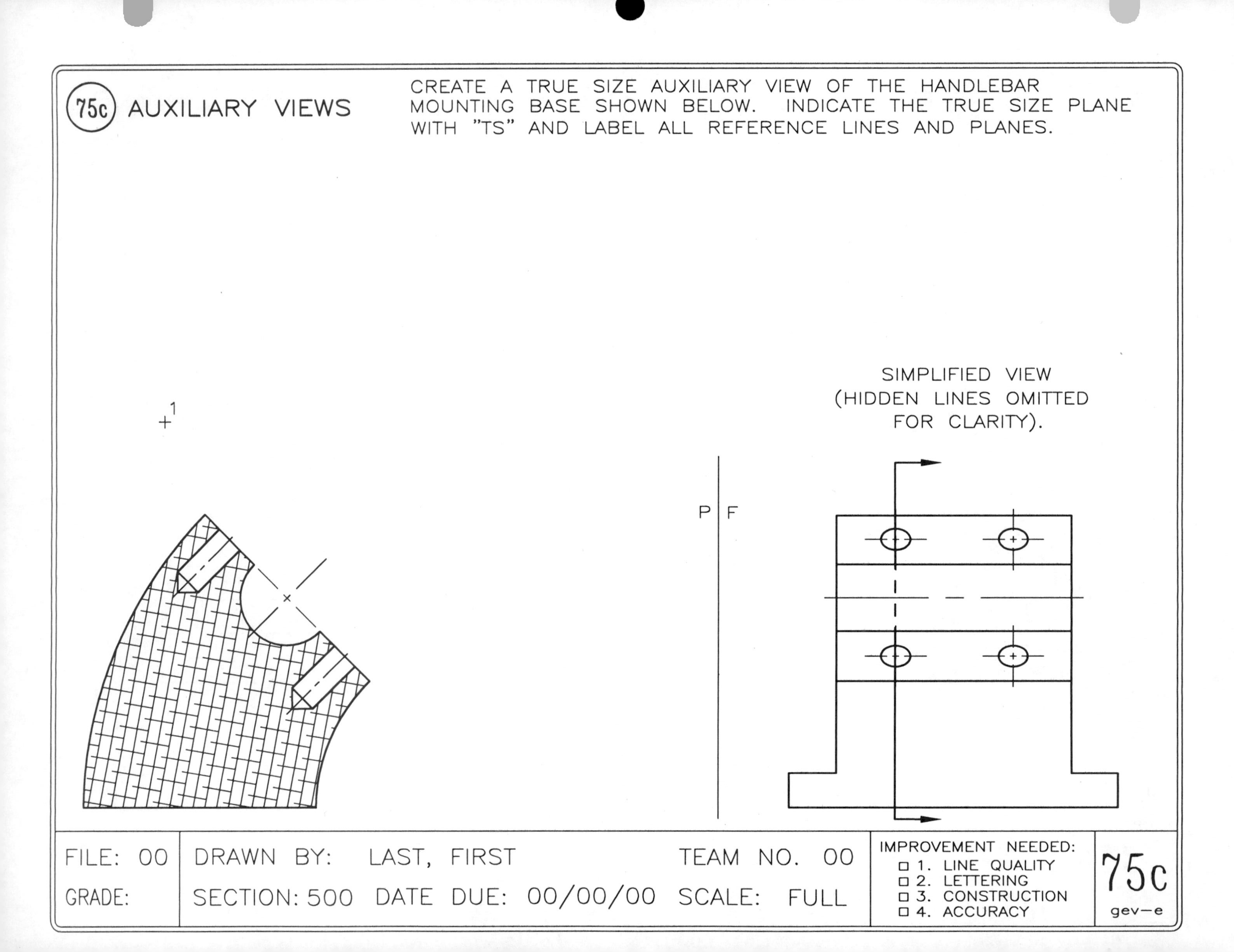

FILE: 00 DRAWN BY: LAST, FIRST TEAM NO. 00

GRADE: SECTION: 500 DATE DUE: 00/00/00 SCALE: FULL

IMPROVEMENT NEEDED:
- ☐ 1. LINE QUALITY
- ☐ 2. LETTERING
- ☐ 3. CONSTRUCTION
- ☐ 4. ACCURACY

75c

gev–e

(76) SOLID MODELS

CREATE SEVERAL PICTORIAL PLANNING SKETCHES OF THE BENCH LEG SHOWN BELOW. ADD NOTES TO YOUR SKETCHES INDICATING THE COMMANDS YOU WILL USE TO CONVERT THEM INTO SOLID MODELS.

(1) WOODEN BENCH LEG

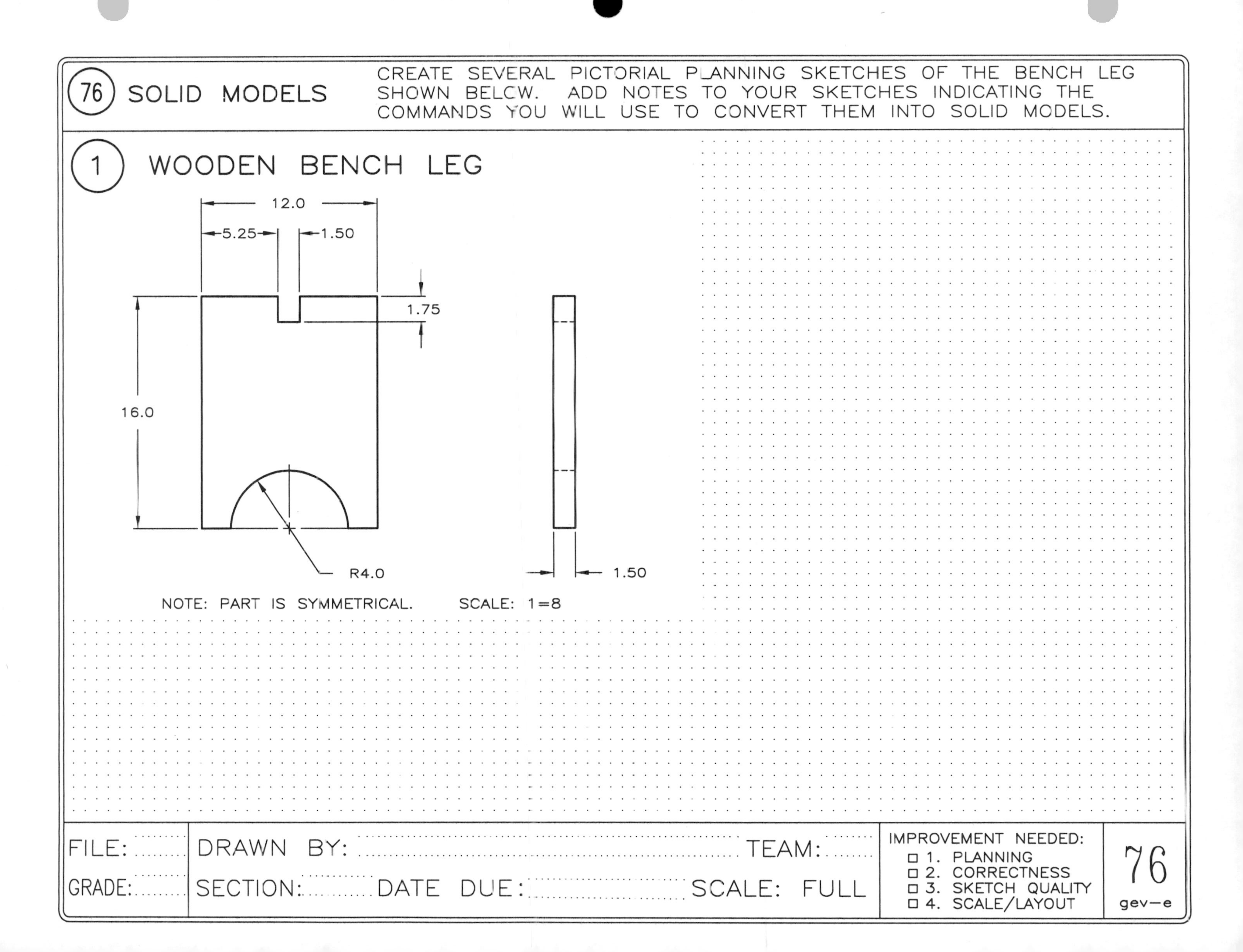

NOTE: PART IS SYMMETRICAL. SCALE: 1=8

FILE:	DRAWN BY:	TEAM:	IMPROVEMENT NEEDED:	76
GRADE:	SECTION: DATE DUE:	SCALE: FULL	☐ 1. PLANNING ☐ 2. CORRECTNESS ☐ 3. SKETCH QUALITY ☐ 4. SCALE/LAYOUT	gev–e

(76c) SOLID MODEL

CREATE A SOLID MODEL OF THE "WOODEN BENCH LEG" FROM THE OTHER SIDE. PLOT FULL SIZE AND HIDDEN FROM A 1,−1,1 VIEWPOINT. DETERMINE THE PROPERTIES OF: MASS, CENTROID, BOUNDING BOX, MOMENTS OF INERTIA, AND PRINT THEM WITH YOUR SOLUTION.

FILE: 00	DRAWN BY: TEAM: 00	IMPROVEMENT NEEDED:	76c
GRADE:.........	SECTION:500 DATE DUE: 00/00/00 SCALE: FULL	□ 1. PEN WIDTHS □ 2. CORRECTNESS □ 3. PLOT QUALITY □ 4. SCALE/LAYOUT	gev−e

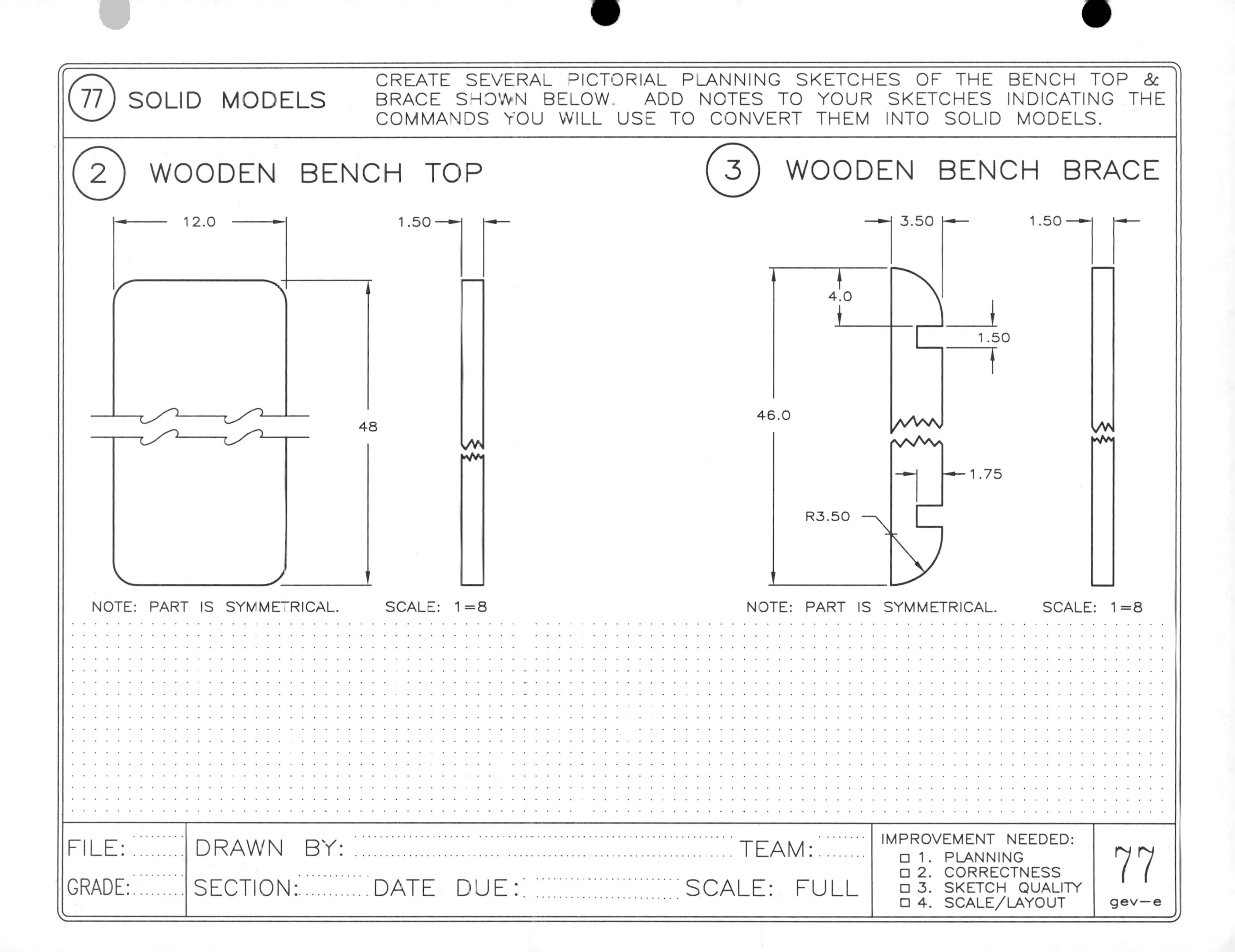
77 SOLID MODELS
CREATE SEVERAL PICTORIAL PLANNING SKETCHES OF THE BENCH TOP & BRACE SHOWN BELOW. ADD NOTES TO YOUR SKETCHES INDICATING THE COMMANDS YOU WILL USE TO CONVERT THEM INTO SOLID MODELS.
2 WOODEN BENCH TOP
12.0
1.50
48
NOTE: PART IS SYMMETRICAL.
SCALE: 1=8
3 WOODEN BENCH BRACE
3.50
1.50
4.0
1.50
46.0
1.75
R3.50
NOTE: PART IS SYMMETRICAL.
SCALE: 1=8
FILE:
GRADE:
DRAWN BY:
TEAM:
SECTION:
DATE DUE:
SCALE: FULL
IMPROVEMENT NEEDED:
1. PLANNING
2. CORRECTNESS
3. SKETCH QUALITY
4. SCALE/LAYOUT
77
gev-e

(77c) SOLID MODEL

CREATE SOLID MODELS OF THE "BENCH TOP & BRACE" FROM THE OTHER SIDE. PLOT FULL SIZE AND HIDDEN FROM A 1,−1,1 VIEWPOINT. DETERMINE THE PROPERTIES OF: MASS, CENTROID, BOUNDING BOX, MOMENTS OF INERTIA, AND PRINT THEM WITH YOUR SOLUTION.

FILE: 00	DRAWN BY: TEAM: 00	IMPROVEMENT NEEDED:	77c
GRADE:.........	SECTION:500 DATE DUE: 00/00/00 SCALE: FULL	☐ 1. LINE WEIGHTS ☐ 2. CORRECTNESS ☐ 3. PLOT QUALITY ☐ 4. SCALE/LAYOUT	gev−e

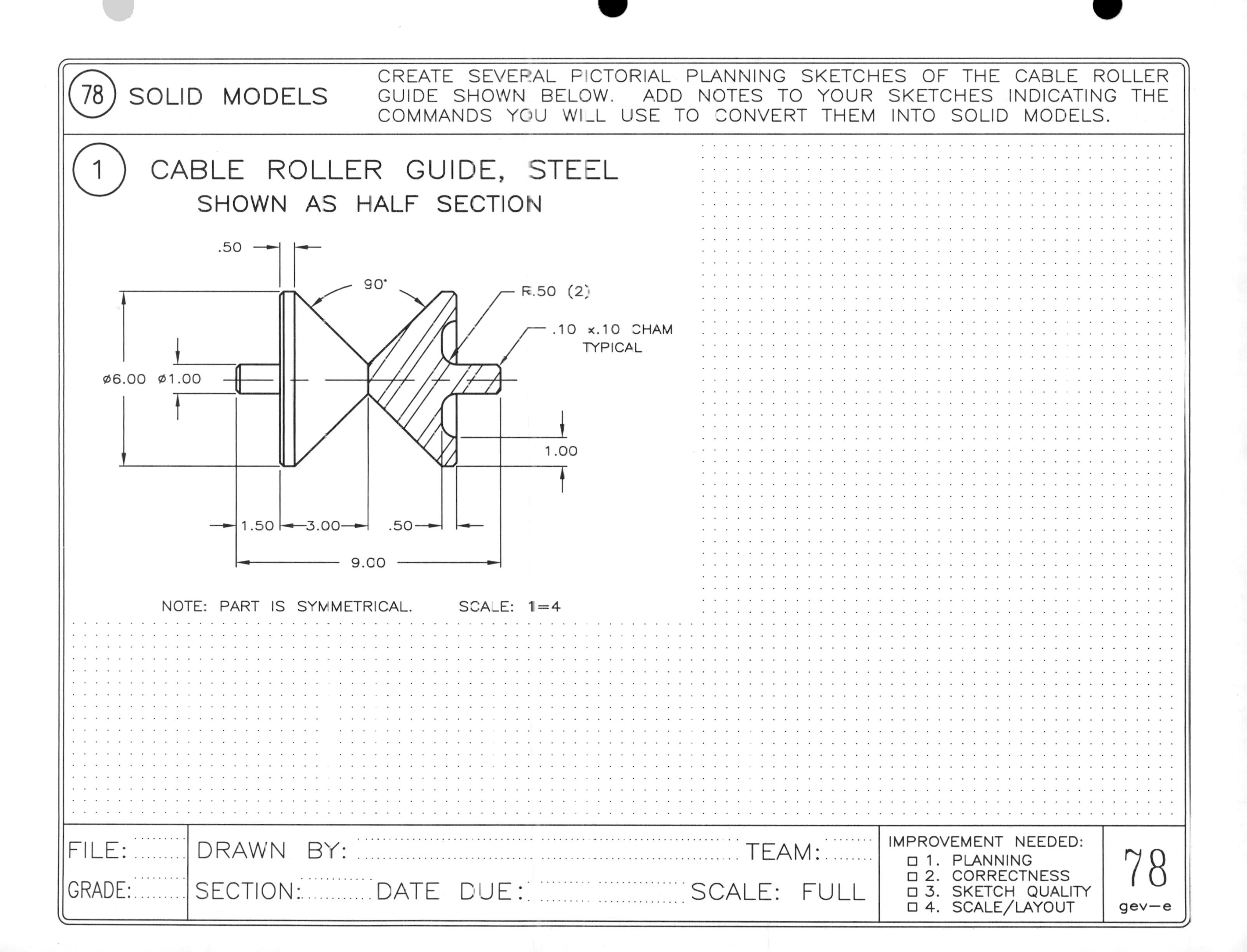

78 SOLID MODELS
CREATE SEVERAL PICTORIAL PLANNING SKETCHES OF THE CABLE ROLLER GUIDE SHOWN BELOW. ADD NOTES TO YOUR SKETCHES INDICATING THE COMMANDS YOU WILL USE TO CONVERT THEM INTO SOLID MODELS.
1 CABLE ROLLER GUIDE, STEEL
SHOWN AS HALF SECTION
.50
90°
R.50 (2)
.10 x.10 CHAM
TYPICAL
ø6.00 ø1.00
1.00
1.50
3.00
.50
9.00
NOTE: PART IS SYMMETRICAL.
SCALE: 1=4
FILE:
GRADE:
DRAWN BY:
TEAM:
SECTION:
DATE DUE:
SCALE: FULL
IMPROVEMENT NEEDED:
☐ 1. PLANNING
☐ 2. CORRECTNESS
☐ 3. SKETCH QUALITY
☐ 4. SCALE/LAYOUT
78
gev-e

(78c) SOLID MODEL

CREATE A SOLID MODEL OF THE "STEEL CABLE ROLLER GUIDE" FROM THE OTHER SIDE. PLOT FULL SIZE AND HIDDEN FROM A 1,−1,1 VIEWPOINT. DETERMINE THE PROPERTIES OF: MASS, CENTROID, BOUNDING BOX, MOMENTS OF INERTIA, AND PRINT THEM WITH YOUR SOLUTION.

FILE: 00

GRADE:.........

DRAWN BY: TEAM: 00

SECTION:500 DATE DUE: 00/00/00 SCALE: FULL

IMPROVEMENT NEEDED:
- □ 1. PEN WIDTHS
- □ 2. CORRECTNESS
- □ 3. PLOT QUALITY
- □ 4. SCALE/LAYOUT

78c

gev−e

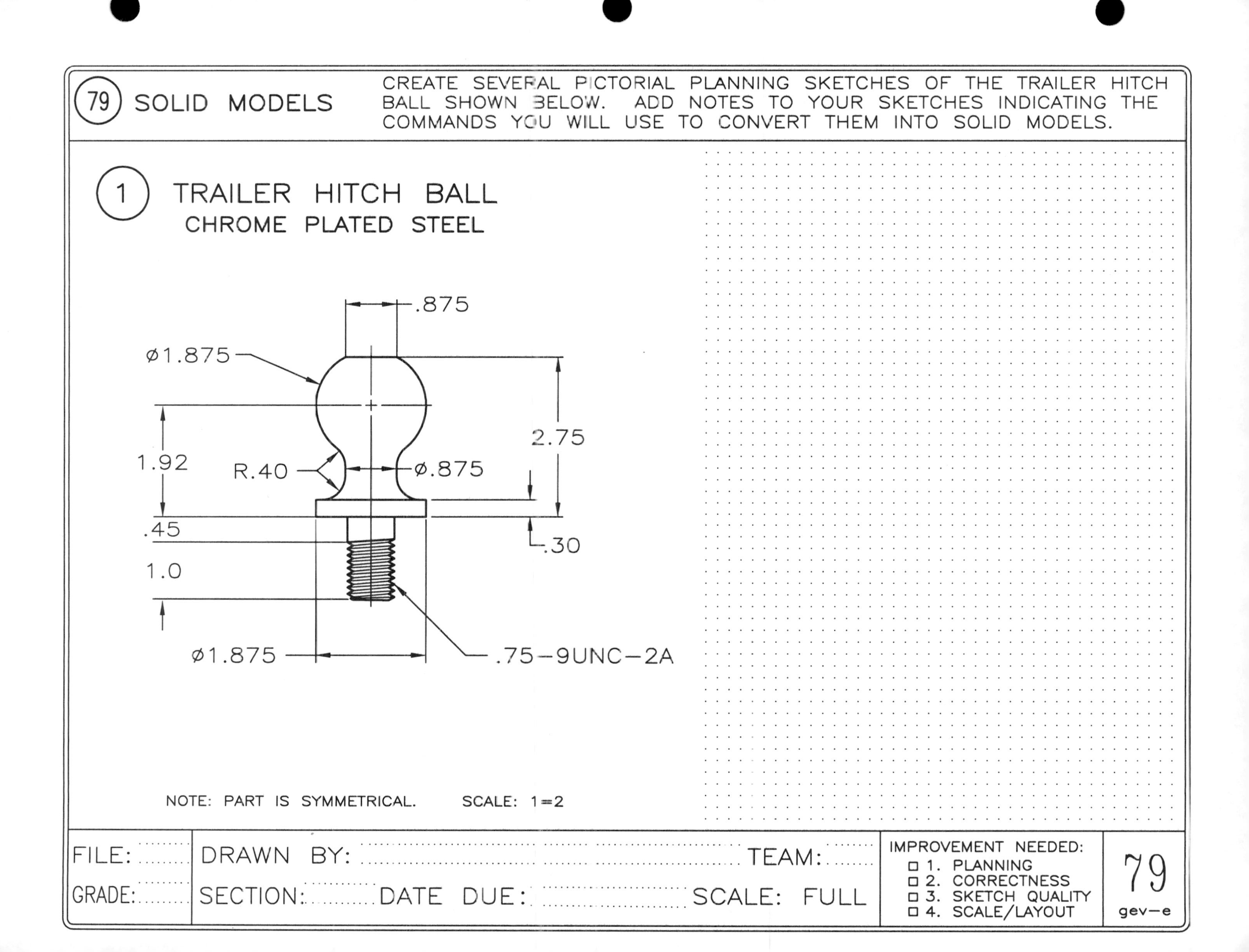

(79) SOLID MODELS
CREATE SEVERAL PICTORIAL PLANNING SKETCHES OF THE TRAILER HITCH BALL SHOWN BELOW. ADD NOTES TO YOUR SKETCHES INDICATING THE COMMANDS YOU WILL USE TO CONVERT THEM INTO SOLID MODELS.
(1) TRAILER HITCH BALL
CHROME PLATED STEEL
.875
⌀1.875
2.75
1.92
R.40
⌀.875
.45
.30
1.0
⌀1.875
.75–9UNC–2A
NOTE: PART IS SYMMETRICAL.
SCALE: 1=2
FILE:
GRADE:
DRAWN BY:
TEAM:
SECTION:
DATE DUE:
SCALE: FULL
IMPROVEMENT NEEDED:
☐ 1. PLANNING
☐ 2. CORRECTNESS
☐ 3. SKETCH QUALITY
☐ 4. SCALE/LAYOUT
79
gev–e

79c SOLID MODEL

CREATE A SOLID MODEL OF THE "STEEL TRAILER HITCH BALL" FROM THE OTHER SIDE. PLOT FULL SIZE AND HIDDEN FROM A 1,−1,1 VIEWPOINT. DETERMINE THE PROPERTIES OF: MASS, CENTROID, BOUNDING BOX, MOMENTS OF INERTIA, AND PRINT THEM WITH YOUR SOLUTION.

FILE: 00	DRAWN BY: TEAM: 00	IMPROVEMENT NEEDED:	79c
GRADE:.........	SECTION:500 DATE DUE: 00/00/00 SCALE: FULL	□ 1. PEN WIDTHS □ 2. CORRECTNESS □ 3. PLOT QUALITY □ 4. SCALE/LAYOUT	gev−e

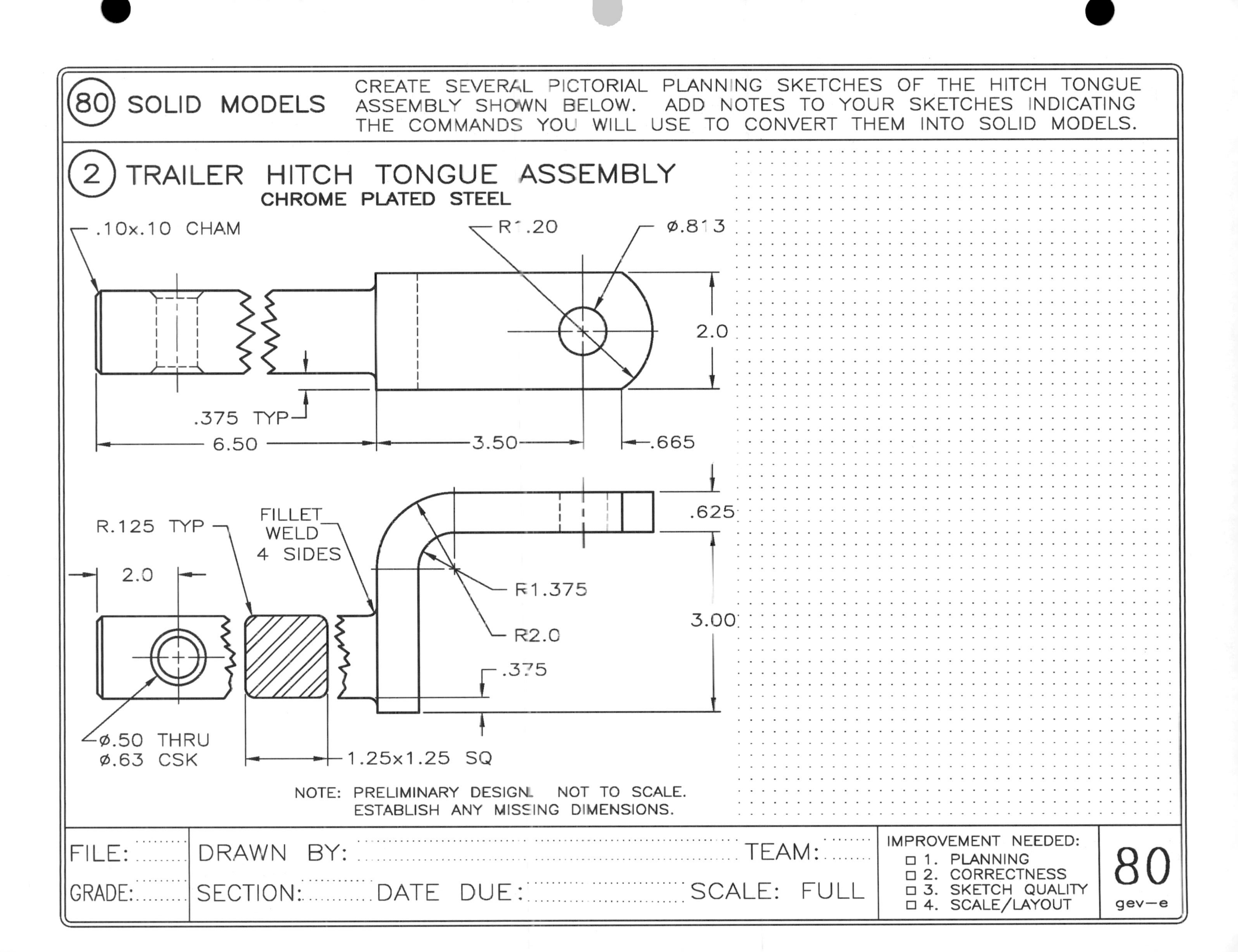
80 SOLID MODELS
CREATE SEVERAL PICTORIAL PLANNING SKETCHES OF THE HITCH TONGUE ASSEMBLY SHOWN BELOW. ADD NOTES TO YOUR SKETCHES INDICATING THE COMMANDS YOU WILL USE TO CONVERT THEM INTO SOLID MODELS.
2 TRAILER HITCH TONGUE ASSEMBLY
CHROME PLATED STEEL
.10x.10 CHAM
R1.20
ø.813
2.0
.375 TYP
6.50
3.50
.665
.625
R.125 TYP
FILLET WELD 4 SIDES
2.0
R1.375
R2.0
3.00
.375
ø.50 THRU
ø.63 CSK
1.25x1.25 SQ
NOTE: PRELIMINARY DESIGN. NOT TO SCALE.
ESTABLISH ANY MISSING DIMENSIONS.
FILE:
GRADE:
DRAWN BY:
TEAM:
SECTION:
DATE DUE:
SCALE: FULL
IMPROVEMENT NEEDED:
☐ 1. PLANNING
☐ 2. CORRECTNESS
☐ 3. SKETCH QUALITY
☐ 4. SCALE/LAYOUT
80
gev-e

(80c) SOLID MODELS

CREATE A SOLID MODEL OF THE "TRAILER HITCH TONGUE ASSEMBLY" FROM THE OTHER SIDE. PLOT FULL SIZE AND HIDDEN FROM A 1,−1,1 VIEWPOINT. DETERMINE THE PROPERTIES OF: MASS, CENTROID, BOUNDING BOX, MOMENTS OF INERTIA, AND PRINT THEM WITH YOUR SOLUTION.

FILE: 00	DRAWN BY: TEAM: 00	IMPROVEMENT NEEDED:	80c
GRADE:.........	SECTION:500 DATE DUE: 00/00/00 SCALE: FULL	☐ 1. LINE WEIGHTS ☐ 2. CORRECTNESS ☐ 3. PLOT QUALITY ☐ 4. SCALE/LAYOUT	gev−e

(81) SOLID MODELS

CREATE SEVERAL PICTORIAL PLANNING SKETCHES OF THE SWIVEL COLLAR SHOWN BELOW. ADD NOTES TO YOUR SKETCHES INDICATING THE COMMANDS YOU WILL USE TO CONVERT THEM INTO SOLID MODELS.

(3) SWIVEL COLLAR, DROP FORGED STEEL

NON-SLIP WAFFLE PATTERN ON TOP MAY VARY. TWO CONCENTRIC CIRCULAR GROOVES AND 14 RADIAL GROOVES ARE USED HERE.

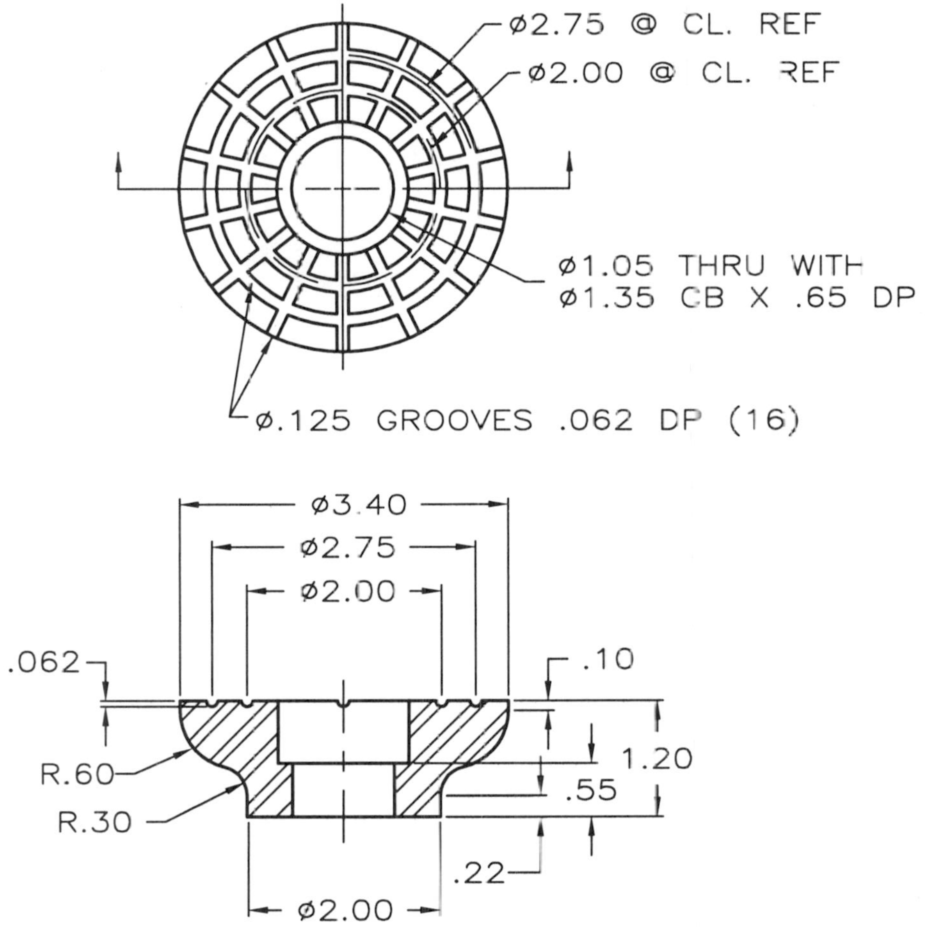

ø2.75 @ CL. REF
ø2.00 @ CL. REF
ø1.05 THRU WITH
ø1.35 CB X .65 DP
ø.125 GROOVES .062 DP (16)

ø3.40
ø2.75
ø2.00
.062
.10
1.20
.55
R.60
R.30
.22
ø2.00

ESTABLISH ANY MISSING DIMENSIONS. SCALE: 1=2

FILE:	DRAWN BY:	TEAM:	IMPROVEMENT NEEDED:	81
GRADE:	SECTION: DATE DUE:	SCALE: FULL	☐ 1. PLANNING ☐ 2. CORRECTNESS ☐ 3. SKETCH QUALITY ☐ 4. SCALE/LAYOUT	gev-e

(81c) SOLID MODELS

CREATE A SOLID MODEL OF THE "STEEL SWIVEL COLLAR" FROM THE OTHER SIDE. PLOT FULL SIZE AND HIDDEN FROM A 1,−1,1 VIEWPOINT. DETERMINE THE PROPERTIES OF: MASS, CENTROID, BOUNDING BOX, MOMENTS OF INERTIA, AND PRINT THEM WITH YOUR SOLUTION.

FILE: 00
GRADE: ....

DRAWN BY: TEAM: 00
SECTION: 500 DATE DUE: 00/00/00 SCALE: FULL

IMPROVEMENT NEEDED:
- ☐ 1. LINE WEIGHTS
- ☐ 2. CORRECTNESS
- ☐ 3. PLOT QUALITY
- ☐ 4. SCALE/LAYOUT

81c
gev–e

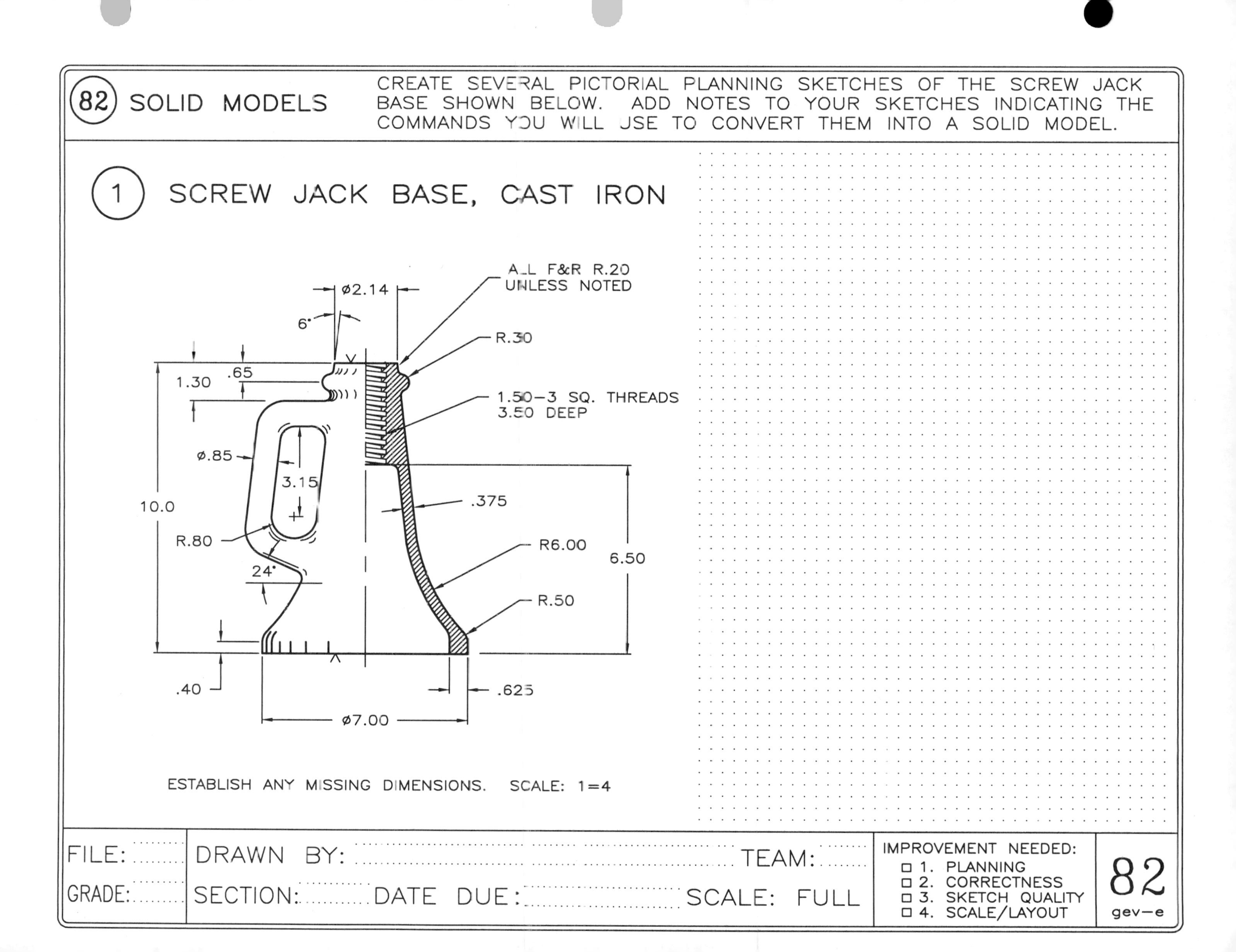
82 SOLID MODELS
CREATE SEVERAL PICTORIAL PLANNING SKETCHES OF THE SCREW JACK BASE SHOWN BELOW. ADD NOTES TO YOUR SKETCHES INDICATING THE COMMANDS YOU WILL USE TO CONVERT THEM INTO A SOLID MODEL.
1 SCREW JACK BASE, CAST IRON
ALL F&R R.20 UNLESS NOTED
ø2.14
6°
R.30
1.30
.65
1.50-3 SQ. THREADS 3.50 DEEP
ø.85
3.15
.375
10.0
R.80
R6.00
6.50
24°
R.50
.40
.625
ø7.00
ESTABLISH ANY MISSING DIMENSIONS. SCALE: 1=4
FILE:
GRADE:
DRAWN BY:
TEAM:
SECTION:
DATE DUE:
SCALE: FULL
IMPROVEMENT NEEDED:
☐ 1. PLANNING
☐ 2. CORRECTNESS
☐ 3. SKETCH QUALITY
☐ 4. SCALE/LAYOUT
82
gev-e

(82c) SOLID MODELS

CREATE A SOLID MODEL OF THE "CAST IRON SCREW JACK BASE" FROM THE OTHER SIDE. PLOT FULL SIZE AND HIDDEN FROM A 1,−1,1 VIEWPOINT. DETERMINE THE PROPERTIES OF: MASS, CENTROID, BOUNDING BOX, MOMENTS OF INERTIA, AND PRINT THEM WITH YOUR SOLUTION.

FILE: 00	DRAWN BY: TEAM: 00	IMPROVEMENT NEEDED:	82c
GRADE:.........	SECTION:500 DATE DUE: 00/00/00 SCALE: FULL	☐ 1. LINE WEIGHTS ☐ 2. CORRECTNESS ☐ 3. PLOT QUALITY ☐ 4. SCALE/LAYOUT	gev−e

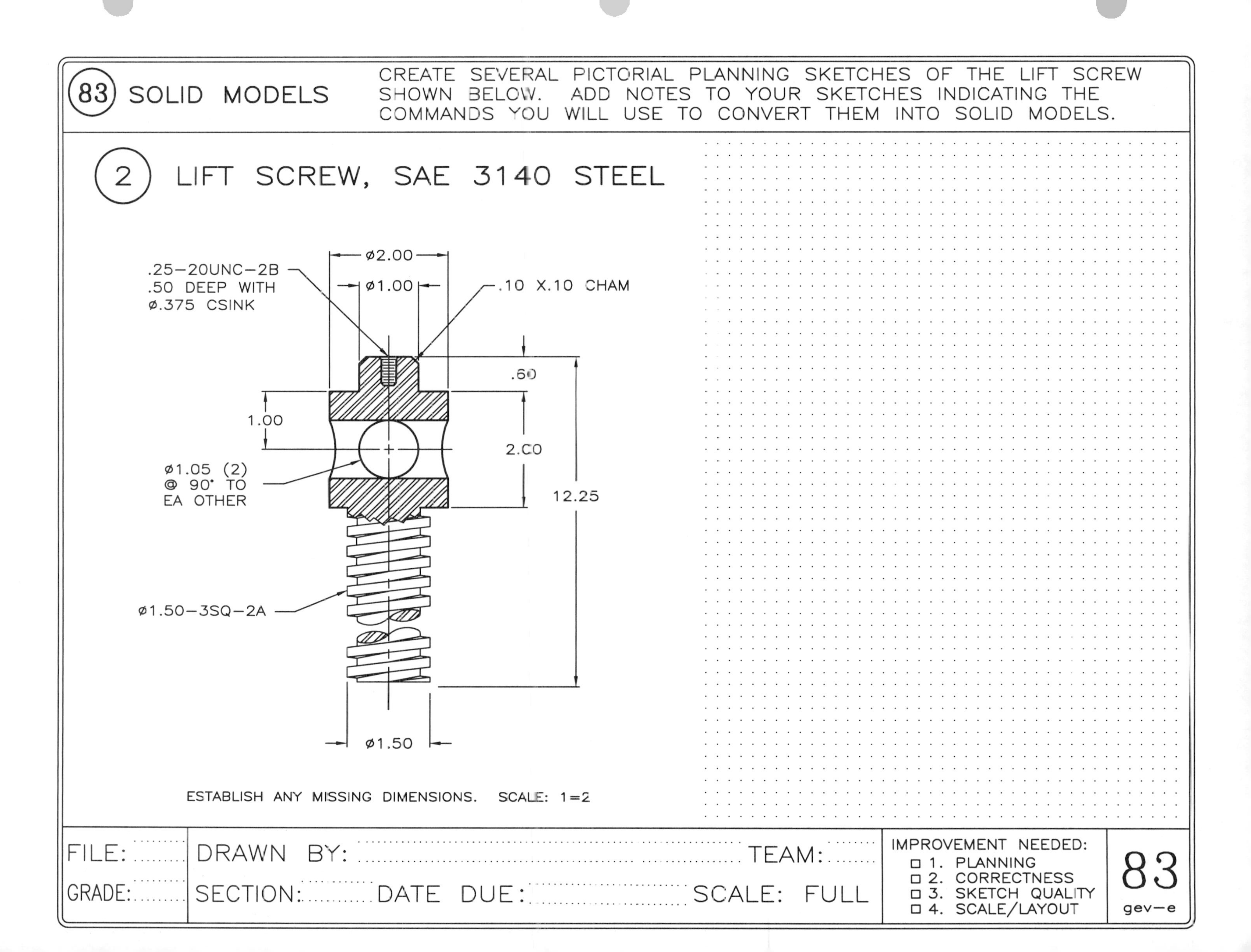
83 SOLID MODELS
CREATE SEVERAL PICTORIAL PLANNING SKETCHES OF THE LIFT SCREW SHOWN BELOW. ADD NOTES TO YOUR SKETCHES INDICATING THE COMMANDS YOU WILL USE TO CONVERT THEM INTO SOLID MODELS.
2 LIFT SCREW, SAE 3140 STEEL
ø2.00
ø1.00
.25–20UNC–2B
.50 DEEP WITH
ø.375 CSINK
.10 X.10 CHAM
.60
1.00
2.00
12.25
ø1.05 (2)
@ 90° TO
EA OTHER
ø1.50–3SQ–2A
ø1.50
ESTABLISH ANY MISSING DIMENSIONS. SCALE: 1=2
FILE:
GRADE:
DRAWN BY:
TEAM:
SECTION:
DATE DUE:
SCALE: FULL
IMPROVEMENT NEEDED:
☐ 1. PLANNING
☐ 2. CORRECTNESS
☐ 3. SKETCH QUALITY
☐ 4. SCALE/LAYOUT
83
gev–e

83c SOLID MODELS

CREATE A SOLID MODEL OF THE "STEEL LIFT SCREW" FROM THE OTHER SIDE. PLOT FULL SIZE AND HIDDEN FROM A 1,−1,1 VIEWPOINT. DETERMINE THE PROPERTIES OF: MASS, CENTROID, BOUNDING BOX, MOMENTS OF INERTIA, AND PRINT THEM WITH YOUR SOLUTION.

FILE: 00	DRAWN BY: TEAM: 00	IMPROVEMENT NEEDED:	83c
GRADE:.........	SECTION:500 DATE DUE: 00/00/00 SCALE: FULL	☐ 1. LINE WEIGHTS ☐ 2. CORRECTNESS ☐ 3. PLOT QUALITY ☐ 4. SCALE/LAYOUT	gev–e

(84) SOLID MODELS

CREATE SEVERAL PICTORIAL PLANNING SKETCHES OF THE ROBOT FOOT SHOWN BELOW. ADD NOTES TO YOUR SKETCHES INDICATING THE COMMANDS YOU WILL USE TO CONVERT THEM INTO SOLID MODELS.

(2) ROBOT FOOT, TITANIUM

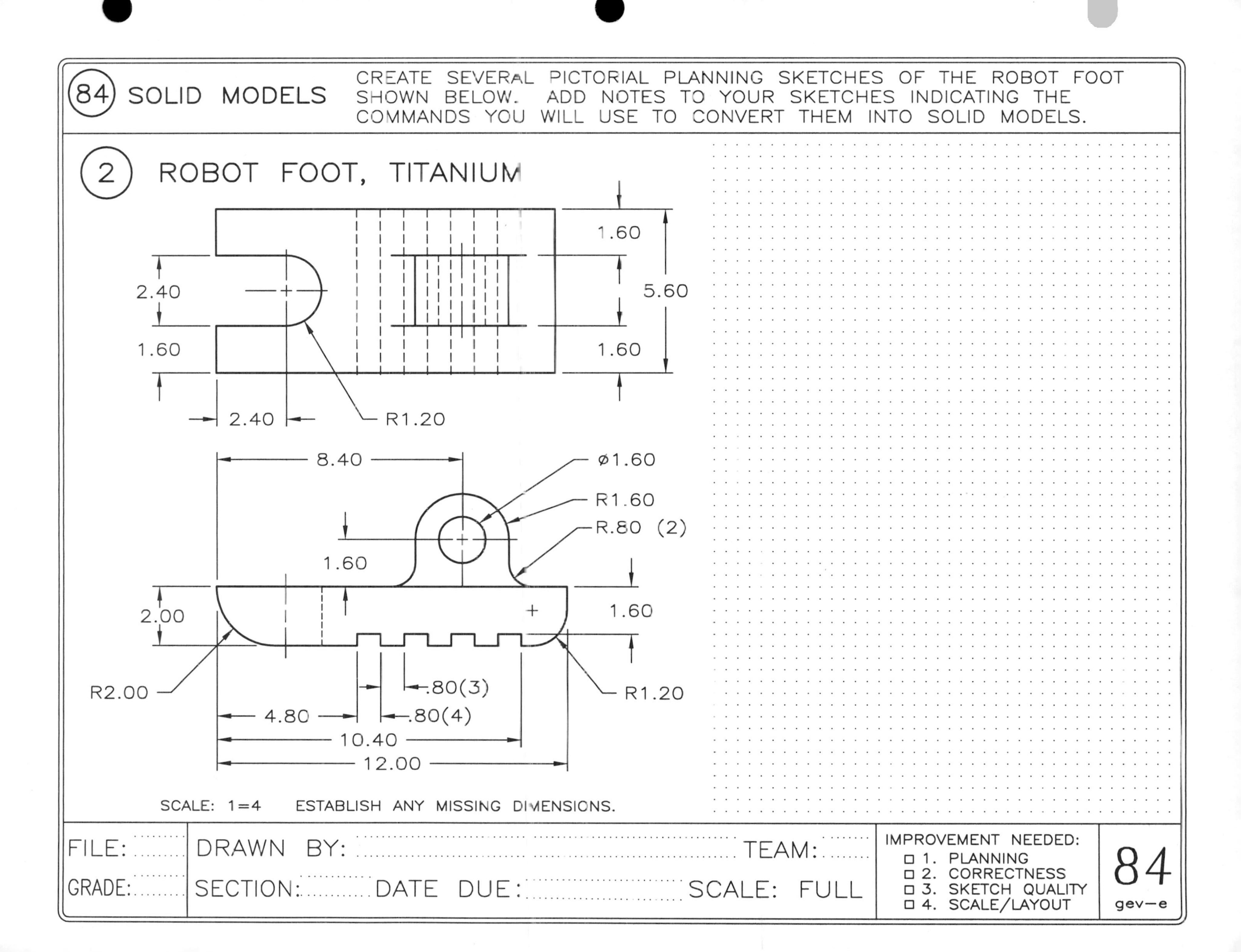

SCALE: 1=4 ESTABLISH ANY MISSING DIMENSIONS.

FILE:	DRAWN BY:	TEAM:	IMPROVEMENT NEEDED:	84
GRADE:	SECTION: DATE DUE:	SCALE: FULL	☐ 1. PLANNING ☐ 2. CORRECTNESS ☐ 3. SKETCH QUALITY ☐ 4. SCALE/LAYOUT	gev–e

84c SOLID MODELS

CREATE A SOLID MODEL OF THE "TITANIUM ROBOT FOOT" FROM THE OTHER SIDE. PLOT FULL SIZE AND HIDDEN FROM A 1,−1,1 VIEWPOINT. DETERMINE THE PROPERTIES OF: MASS, CENTROID, BOUNDING BOX, MOMENTS OF INERTIA, AND PRINT THEM WITH YOUR SOLUTION.

FILE: 00	DRAWN BY: TEAM: 00	IMPROVEMENT NEEDED:	84c
GRADE:.........	SECTION:500 DATE DUE: 00/00/00 SCALE: FULL	☐ 1. LINE WEIGHTS ☐ 2. CORRECTNESS ☐ 3. PLOT QUALITY ☐ 4. SCALE/LAYOUT	gev−e

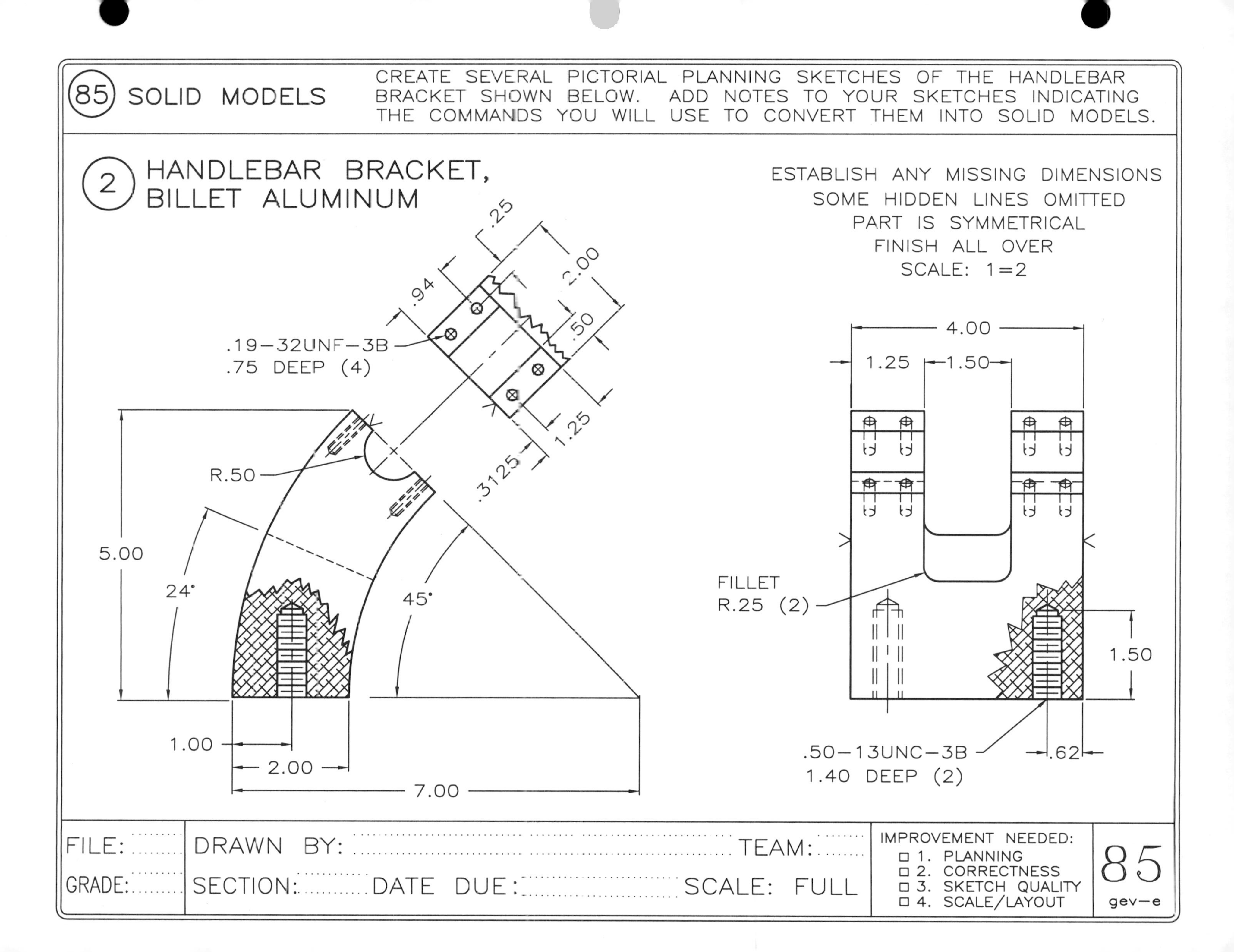

85 SOLID MODELS
CREATE SEVERAL PICTORIAL PLANNING SKETCHES OF THE HANDLEBAR BRACKET SHOWN BELOW. ADD NOTES TO YOUR SKETCHES INDICATING THE COMMANDS YOU WILL USE TO CONVERT THEM INTO SOLID MODELS.
2 HANDLEBAR BRACKET, BILLET ALUMINUM
ESTABLISH ANY MISSING DIMENSIONS
SOME HIDDEN LINES OMITTED
PART IS SYMMETRICAL
FINISH ALL OVER
SCALE: 1=2
.25
.94
2.00
.50
1.25
.3125
.19–32UNF–3B
.75 DEEP (4)
R.50
5.00
24°
45°
1.00
2.00
7.00
4.00
1.25
1.50
FILLET
R.25 (2)
1.50
.50–13UNC–3B
1.40 DEEP (2)
.62
FILE:
GRADE:
DRAWN BY:
TEAM:
SECTION:
DATE DUE:
SCALE: FULL
IMPROVEMENT NEEDED:
☐ 1. PLANNING
☐ 2. CORRECTNESS
☐ 3. SKETCH QUALITY
☐ 4. SCALE/LAYOUT
85
gev–e

85c SOLID MODELS

CREATE A SOLID MODEL OF THE "ALUMINUM HANDLEBAR BRACKET" FROM THE OTHER SIDE. PLOT FULL SIZE AND HIDDEN FROM A 1,−1,1 VIEWPOINT. DETERMINE THE PROPERTIES OF: MASS, CENTROID, BOUNDING BOX, MOMENTS OF INERTIA, AND PRINT THEM WITH YOUR SOLUTION.

FILE: 00 GRADE:	DRAWN BY: TEAM: 00 SECTION:500 DATE DUE: 00/00/00 SCALE: FULL	IMPROVEMENT NEEDED: ☐ 1. LINE WEIGHTS ☐ 2. CORRECTNESS ☐ 3. PLOT QUALITY ☐ 4. SCALE/LAYOUT	85c gev−e

DESIGN TEAM PROJECT

PROJECT TITLE:

....
....

DESIGN TEAM NAME:

....
....

DESIGN TEAM MEMBERS:

	NAME	FILE	% PARTICIPATION
1.	....	....	....
2.	....	....	....
3.	....	....	....
4.	....	....	....
5.	....	....	....
6.	....	....	....
7.	....	....	....
8.	....	....	....

NAME: .... GRADE: ....

FILE: .... DATE: .... SECTION: ....

PROBLEM IDENTIFICATION*

PRINT NEATLY WITH ALL CAPS, SINGLE STROKE GOTHIC.

1. PROJECT TITLE:

2. PROBLEM STATEMENT:

3. REQUIREMENTS AND LIMITATIONS:

* REFER TO "ESSENTIALS OF ENGINEERING GRAPHICS" 1st Ed. 2002, BY GERALD E. VINSON FOR EXAMPLES OF THE TEAM DESIGN FORMAT FEATURED IN THIS WORKBOOK.

FILE:	NAME:		TEAM:	SECT:	DWG NO.
	GRADE:	DATE:			DP1

PROBLEM IDENTIFICATION

4. NEEDED INFORMATION

5. MARKET CONSIDERATIONS

FILE:	NAME:		TEAM:	SECT:	DWG NO.
	GRADE:	DATE:			DP2

PRELIMINARY IDEAS

1. BRAINSTORMING RESULTS:

FILE:	NAME:		TEAM:	SECT:	DWG NO.
	GRADE:	DATE:			DP3

PRELIMINARY IDEAS

2. SKETCHES & DESCRIPTION OF BEST IDEAS:

FILE:	NAME:		TEAM:	SECT:	DWG NO.
	GRADE:	DATE:			DP4

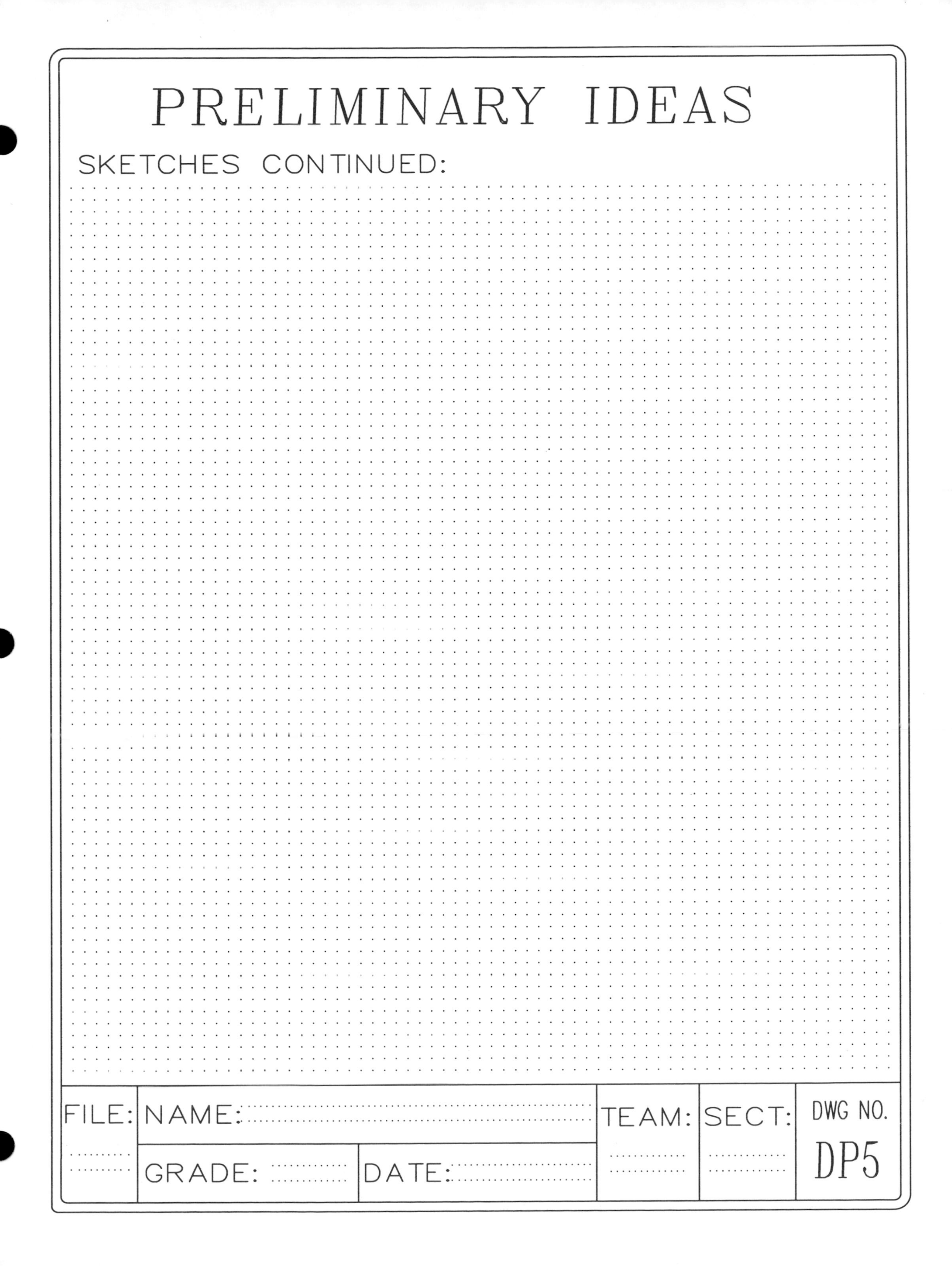
PRELIMINARY IDEAS
SKETCHES CONTINUED:
FILE:
NAME:
TEAM:
SECT:
DWG NO.
DP5
GRADE:
DATE:

PRELIMINARY IDEAS

SKETCHES CONTINUED:

FILE:	NAME:		TEAM:	SECT:	DWG NO.
	GRADE:	DATE:			DP6

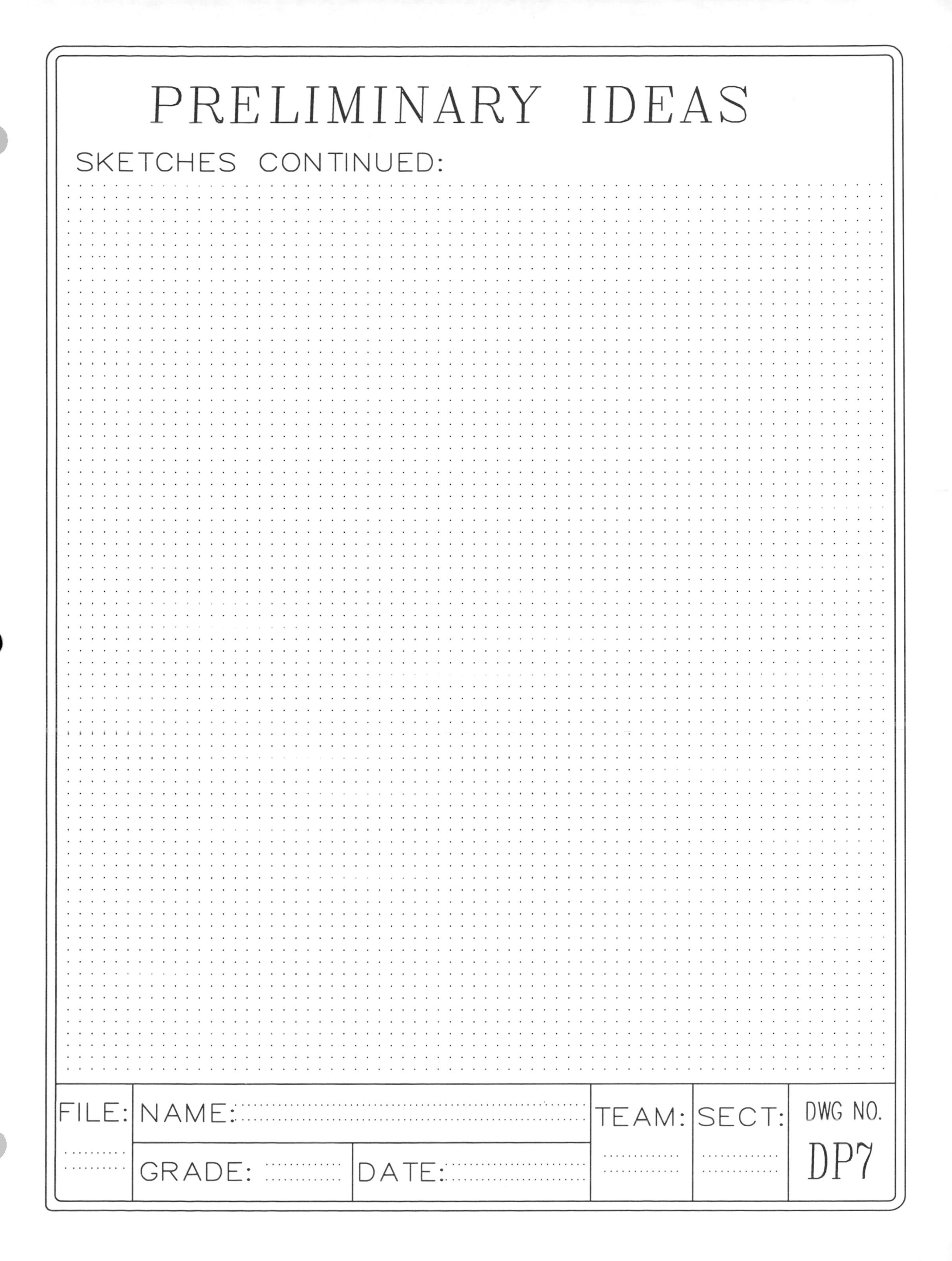
PRELIMINARY IDEAS
SKETCHES CONTINUED:
FILE:
NAME:
TEAM:
SECT:
DWG NO.
DP7
GRADE:
DATE:

REFINEMENT

1. DESCRIPTION OF BEST IDEA:

2. ATTACH SCALE DRAWINGS:

FILE:	NAME:		TEAM:	SECT:	DWG NO.
	GRADE:	DATE:			DP8

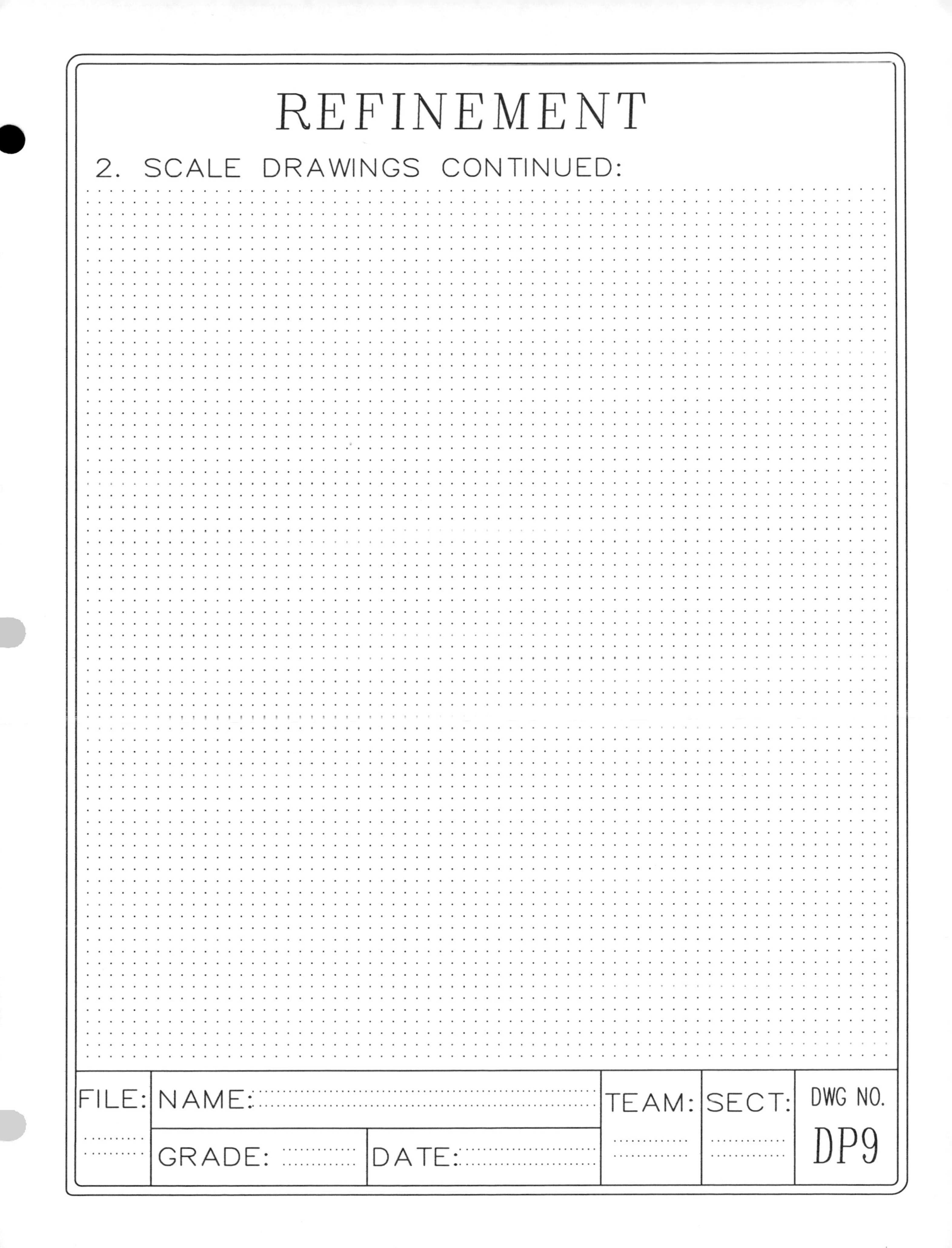
REFINEMENT
2. SCALE DRAWINGS CONTINUED:
FILE:
NAME:
GRADE:
DATE:
TEAM:
SECT:
DWG NO.
DP9

ANALYSIS

1. WHAT IS THE FUNCTION OF THIS DESIGN:

2. HUMAN FACTORS ENGINEERING:

3. MARKET AND CONSUMER ACCEPTANCE:

(ATTACH GRAPHS)

FILE:	NAME:		TEAM:	SECT:	DWG NO. DP10
	GRADE:	DATE:			

ANALYSIS

4. DETAILED PHYSICAL DESCRIPTION:

5. STRENGTH VERIFICATION:

FILE:	NAME:		TEAM:	SECT:	DWG NO.
	GRADE:	DATE:			DP11

ANALYSIS

6. MANUFACTURING PROCESSES:

7. ECONOMIC FACTORS:

FILE:	NAME:		TEAM:	SECT:	DWG NO.
	GRADE:	DATE:			DP12

DECISION

1. DESIGN EVALUATION SUMMARY TABLE:

A. DESIGN 1: ....

B. DESIGN 2: ....

C. DESIGN 3: ....

D. DESIGN 4: ....

E. DESIGN 5: ....

F. DESIGN 6: ....

G. DESIGN 7: ....

MAX VALUE	FACTORS TO ANALYZE	DESIGNS 1	2	3	4	5	6	7
	FUNCTION							
	HUMAN FACTORS							
	MARKET ANALYSIS							
	STRENGTH							
	PRODUCTION PRO.							
	COST ANALYSIS							
	PROFITABILITY							
	APPEARANCE							
100	TOTALS:							

FILE:	NAME:		TEAM:	SECT:	DWG NO. DP13
	GRADE:	DATE:			

CONCLUSIONS:

FILE:	NAME:		TEAM:	SECT:	DWG NO.
	GRADE:	DATE:			DP14

IMPLEMENTATION

FINAL SCALE DRAWINGS:

FILE:	NAME:		TEAM:	SECT:	DWG NO.
	GRADE:	DATE:			DP15

DRAWING WORK SHEET

FILE:	NAME:		TEAM:	SECT:	DWG NO.
	GRADE:	DATE:			

ISO .10

FILE:	GRADE:	IMPROVEMENT NEEDED:	NAME:	SECTION:	DWG.
		☐ 1. LINE QUALITY ☐ 2. LETTERING ☐ 3. CONSTRUCTION ☐ 4. ACCURACY	DATE: SCALE:		gev–e

ISO .10

FILE:	GRADE:	IMPROVEMENT NEEDED: ☐ 1. LINE QUALITY ☐ 2. LETTERING ☐ 3. CONSTRUCTION ☐ 4. ACCURACY	NAME: DATE: SCALE:	SECTION:	DWG.

gev–e

ISO .10

FILE:

GRADE:

IMPROVEMENT NEEDED:
- □ 1. LINE QUALITY
- □ 2. LETTERING
- □ 3. CONSTRUCTION
- □ 4. ACCURACY

NAME:

DATE:

SCALE:

SECTION:

DWG.

gev-e

ISO .10

FILE:

GRADE:

IMPROVEMENT NEEDED:
- ☐ 1. LINE QUALITY
- ☐ 2. LETTERING
- ☐ 3. CONSTRUCTION
- ☐ 4. ACCURACY

NAME:

DATE: SCALE:

SECTION:

DWG.

gev–e

ISO .10

FILE:	GRADE:	IMPROVEMENT NEEDED: ☐ 1. LINE QUALITY ☐ 2. LETTERING ☐ 3. CONSTRUCTION ☐ 4. ACCURACY	NAME: DATE: SCALE:	SECTION:	DWG. gev–e

ISO .10

FILE:	GRADE:	IMPROVEMENT NEEDED: ☐ 1. LINE QUALITY ☐ 2. LETTERING ☐ 3. CONSTRUCTION ☐ 4. ACCURACY	NAME: DATE: SCALE:	SECTION:	DWG. gev–e

ISO .10

FILE:

GRADE:

IMPROVEMENT NEEDED:
- ☐ 1. LINE QUALITY
- ☐ 2. LETTERING
- ☐ 3. CONSTRUCTION
- ☐ 4. ACCURACY

NAME:

DATE:

SCALE:

SECTION:

DWG.

gev–e

ISO .10

FILE:	GRADE:	IMPROVEMENT NEEDED: ☐ 1. LINE QUALITY ☐ 2. LETTERING ☐ 3. CONSTRUCTION ☐ 4. ACCURACY	NAME: DATE: SCALE:	SECTION:	DWG. gev–e

SQ .10

FILE:	GRADE:	IMPROVEMENT NEEDED: ☐ 1. LINE QUALITY ☐ 2. LETTERING ☐ 3. CONSTRUCTION ☐ 4. ACCURACY	NAME: DATE: SCALE:	SECTION:	DWG.

gev–e

SQ .10

FILE:

GRADE:

IMPROVEMENT NEEDED:
- ☐ 1. LINE QUALITY
- ☐ 2. LETTERING
- ☐ 3. CONSTRUCTION
- ☐ 4. ACCURACY

NAME:

DATE:

SCALE:

SECTION:

DWG.

gev-e

SQ .10

FILE:	GRADE:	IMPROVEMENT NEEDED: ☐ 1. LINE QUALITY ☐ 2. LETTERING ☐ 3. CONSTRUCTION ☐ 4. ACCURACY	NAME: DATE: SCALE:	SECTION:	DWG.

gev-e

SQ .10

FILE:	GRADE:	IMPROVEMENT NEEDED: ☐ 1. LINE QUALITY ☐ 2. LETTERING ☐ 3. CONSTRUCTION ☐ 4. ACCURACY	NAME: DATE: SCALE:	SECTION:	DWG. gev–e

SQ .10

FILE:

GRADE:

IMPROVEMENT NEEDED:
- ☐ 1. LINE QUALITY
- ☐ 2. LETTERING
- ☐ 3. CONSTRUCTION
- ☐ 4. ACCURACY

NAME:

DATE:

SCALE:

SECTION:

DWG.

gev–e

SQ .10

FILE:	GRADE:	IMPROVEMENT NEEDED: ☐ 1. LINE QUALITY ☐ 2. LETTERING ☐ 3. CONSTRUCTION ☐ 4. ACCURACY	NAME: DATE: SCALE:	SECTION:	DWG. gev–e

SQ .10

FILE:	GRADE:	IMPROVEMENT NEEDED: ☐ 1. LINE QUALITY ☐ 2. LETTERING ☐ 3. CONSTRUCTION ☐ 4. ACCURACY	NAME: DATE: SCALE:	SECTION:	DWG.

gev–e

SQ .10

FILE:

GRADE:

IMPROVEMENT NEEDED:
- ☐ 1. LINE QUALITY
- ☐ 2. LETTERING
- ☐ 3. CONSTRUCTION
- ☐ 4. ACCURACY

NAME:

DATE: SCALE:

SECTION:

DWG.

gev–e

SQ .10

FILE:	GRADE:	IMPROVEMENT NEEDED: ☐ 1. LINE QUALITY ☐ 2. LETTERING ☐ 3. CONSTRUCTION ☐ 4. ACCURACY	NAME: DATE: SCALE:	SECTION:	DWG.

gev-e

SQ .10

FILE:	GRADE:	IMPROVEMENT NEEDED: ☐ 1. LINE QUALITY ☐ 2. LETTERING ☐ 3. CONSTRUCTION ☐ 4. ACCURACY	NAME: DATE: SCALE:	SECTION:	DWG. gev-e

WD

FILE:	GRADE:	DRAWN BY: CHECKED BY: APPROVED BY:	NAME: SECT: DATE: SCALE:	TOLERANCES: LIN: ANG:	PAGE OF gev-e

WD

FILE:	GRADE:	DRAWN BY: CHECKED BY: APPROVED BY:	NAME: SECT: DATE: SCALE:	TOLERANCES: LIN: ANG:	PAGE OF gev–e

WD

FILE:	GRADE:	DRAWN BY: CHECKED BY: APPROVED BY:	NAME: SECT: DATE: SCALE:	TOLERANCES: LIN: ANG:	PAGE OF gev-e

WD

FILE:	GRADE:	DRAWN BY: CHECKED BY: APPROVED BY:	NAME: SECT: DATE: SCALE:	TOLERANCES: LIN: ANG:	PAGE OF gev–e

WD

FILE:	GRADE:	DRAWN BY: CHECKED BY: APPROVED BY:	NAME: SECT: DATE: SCALE:	TOLERANCES: LIN: ANG:	PAGE OF gev–e

WD

FILE:	GRADE:	DRAWN BY: CHECKED BY: APPROVED BY:	NAME: SECT: DATE: SCALE:	TOLERANCES: LIN: ANG:	PAGE OF gev—e

APPENDIX 1: USEFUL AutoCAD® COMMAND ALIASES

A ....ARC
AR..................ARRAY
BH..................BHATCH
BR..................BREAK
C..................CIRCLE
CH..................DDCHPROP
CHA..................CHAMFER
CO..................COPY
DED..................DIMEDIT
DI..................DIST
DIV..................DIVIDE
DO..................DONUT
DRA..................DIMRADIUS
DST..................DIMSTYLE
DT..................DTEXT
DV..................DVIEW
E..................ERASE
ED..................DDEDIT
EL..................ELLIPSE
EX..................EXTEND
F..................FILLET
H..................BHATCH
HE..................HATCHEDIT
HI..................HIDE
I..................DDINSERT
IN..................INTERSECT
IO..................INSERTOBJ
L..................LINE
LA..................LAYER
LE..................LEADER
LEN..................LENGTHEN
LS..................LIST
LT..................LINETYPE
LTS..................LTSCALE
M..................MOVE
MA..................MATCHPROP
MI..................MIRROR
ML..................MLINE
MO..................DDMODIFY
MS..................MSPACE
MT..................MTEXT
MV..................MVIEW
O..................OFFSET
P..................PAN
PE..................PEDIT
PL..................PLINE
PO..................POINT
POL..................POLYGON
PR..................PREFERENCES
PLOT..................PRINT
PS..................PSPACE
PU..................PURGE
R..................REDRAW
RA..................REDRAWALL
RE..................REGEN
REA..................REGENALL
REC..................RECTANGLE
REG..................REGION
REV..................REVOLVE
RO..................ROTATE
S..................STRETCH
SC..................SCALE
SEC..................SECTION
SET..................SETVAR
SL..................SLICE
SN..................SNAP
SP..................SPELL
SPL..................SPLINE
SPE..................SPLINEDIT
ST..................STYLE
SU..................SUBTRACT
T..................MTEXT
TA..................TABLET
TH..................THICKNESS
TI..................TILEMODE
TOL..................TOLERANCE
TOR..................TORUS
TR..................TRIM
UC..................DDUCS
UN..................DDUNITS
UNI..................UNION
V..................DDVIEW
VP..................DDVPOINT
W..................WBLOCK
WE..................WEDGE
X..................EXPLODE
XA..................XATTACH
XL..................XLINE
XR..................XREF
Z..................ZOOM

APPENDIX 2: USEFUL ABBREVIATIONS (ANSI Y1.1)

Actual ACT
Adapter ADPT
Allowance ALLOW
Alloy ALY
Aluminum AL
American National Standards Institute ANSI
Amount AMT
Anneal ANL
Apparatus APP
Appendix APPX
Approved APPD
Approximate APPROX
Assembly ASSY
Association ASSN
Authorized AUTH
Auxiliary AUX
Average AVG
Babbitt BAB
Back pressure BP
Ball Bearing BB
Base Line BL
Bearing BRG
Between BET
Between centers BC
Bevel BEV
Bill of material BOM
Board BD
Bolt circle BC
Both sides BS
Bottom BOT
Brass BRS
Break BRK
Brinnell hardness BH
British Thermal Units BTU
Broach BRO
Bronze BRZ
Building BLDG
Bushing BUSH.
Cadmium plate CD PL
Calibrate CAL
Cap screw CAP SCR
Case harden CH
Cast iron CI
Cast steel CS
Castle nut CAS NUT
Center CTR
Centerline CL
Center to center C to C
Chain CH
Chamfer CHAM
Change notice CN
Change order CO
Channel CHAN
Check CHK
Chord CHD
Circle CIR
Circumference CIRC
Clockwise CW
Coated CTD
Cold drawn steel CDS
Cold rolled steel CRS
Column COL
Combination COMB
Company CO
Concentric CONC
Condition COND
Connect CONN
Cotter COT
Counterclockwise CCW
Counterbore CBORE
Counterdrill CDRILL
Countersink CSK
Coupling CPLG
Crank CRK
Cross section XSECT
Cubic foot CU FT
Cubic inch CU IN
Cylinder CYL
Decimal DEC
Degree (°) DEG
Department DEPT
Design DSGN
Detail DET
Diagonal DIAG
Diameter DIA
Diametrical pitch DP
Dimension DIM
Distance DIST
Dovetail DVTL
Down DN
Drafting DFTG
Drawing DWG
Drill DR
Drop forge DF
Each EA
Eccentric ECC
Effective EFF
Elbow ELL
Elevation ELEV
Engineer ENGR
Equal EQ
Equipment EQUIP
Equivalent EQUIV
Exterior EXT
Extra heavy X HVY
Extra strong X STR
Fabricate FAB
Face to face F to F
Fahrenheit F
Far side FS
Federal FED
Feet (′) FT
Feet per minute FPM
Feet per second FPS
Figure FIG
Fillet FIL
Fillister FIL
Finish FIN
Finish all over FAO
Flange FLG
Flat head FH
Foot (′) FT
Forging FORG
Frequency FREQ
Front FR
Gage GA
Galvanize GALV
Galvanized iron GI
Galvanized steel GS
Gasket GSKT
General GEN
Grade GR
Grind GRD
Groove GRV
Ground GRD
Harden HDN
Head HD
Headquarters HQ
Heat treat HT TR
Hexagon HEX
High-speed HS
Horizontal HOR
Hot rolled steel HRS
Hydraulic HYD
Illustrate ILLUS
Inch (″) IN.
Inches per second IPS
Include INCL
Industrial IND
Information INFO
Inside diameter ID
Insulate INS
Interior INT
Internal INT
Intersect INT

Iron I
Junction .... JCT
Key .... K
Keyseat .... KST
Keyway .... KWY
Kip (1000 lb) .... K
Knots .... KN
Laboratory .... LAB
Lateral .... LAT
Left L
Left hand .... LH
Letter .... LTR
Line L
Lubricate .... LUB
Machine .... MACH
Maileable iron .... MI
Manufacture .... MFR
Material .... MATL
Maximum .... MAX
Mechanical .... MECH
Metal .... MET
Millimeter .... MM
Minimum .... MIN
Minute .... (") MIN
Miscellaneous .... MISC
Model .... MOD
Month .... MO
Morse taper .... MOR T
Multiple .... MULT
National .... NATL
Negative .... NEG
Nipple .... NIP
Nominal .... NOM
Normal .... NOR
Not to scale .... NTS
Number .... NO.
Obsolete .... OBS
On center .... OC
Opposite .... OPP
Original .... ORIG
Outside diameter .... OD
Outside face .... OF
Overall .... OA
Parallel .... PAR.
Permanent .... PERM
Perpendicular .... PERP
Piece .... PC
Pitch .... P
Pitch circle .... PC
Pitch diameter .... PD
Plastic .... PLSTC
Plate .... PL
Point .... PT
Pound .... LB
Power .... PWR
Prefabricated .... PREFAB
Preferred .... PFD

Irregular .... IRREG
Pressure .... PRESS
Production .... PROD
Project .... PROJ
Quadrant .... QUAD
Quarter .... QTR
Radius .... R
Ream .... RM
Received .... RECD
Rectangle .... RECT
Reference .... REF
Remove .... REM
Require .... REQ
Required .... REQD
Revise .... REV
Revolution .... REV
Right .... R
Right hand .... RH
Rivet .... RIV
Rockwell hardness .... RH
Roller bearing .... RB
Root diameter .... RD
Rough .... RGH
Round .... RD
Rubber .... RUB.
Schedule .... SCH
Screw .... SCR
Second .... SEC
Section .... SECT
Separate .... SEP
Set screw .... SS
Shaft .... SFT
Sketch .... SK
Sleeve .... SLV
Slotted .... SLOT
Socket .... SOC
Space .... SP
Special .... SPL
Spherical .... SPHER
Spot faced .... SF
Spring .... SPG
Square .... SQ
Stainless steel .... SST
Standard .... STD
Steel .... STL
Stock .... STK
Straight .... STR
Substitute .... SUB
Symmetrical .... SYM
Tangent .... TAN.
Taper .... TPR
Technical .... TECH
Temperature .... TEMP
Template .... TEMP
Tension .... TENS
Thick .... THK
Thread .... THD

Joint .... JT
Tolerance .... TOL
Tool steel .... TS
Total .... TOT
Typical .... TYP
Ultimate .... ULT
Unit .... U
Universal .... UNIV
Vacuum .... VAC
Variable .... VAR
Vertical .... VERT
Volume .... VOL
Washer .... WASH
Weight .... WT
Width .... W
Wood .... WD
Woodruff .... WDF
Wrought iron .... WI
Yard .... YD
Year .... YR

COLORS:

Amber .... AMB
Black .... BLK
Blue .... BLU
Brown .... BRN
Green .... GRN
Orange .... ORN
White .... WHT
Yellow .... YEL

AutoCAD® COLOR CODES:

Red 1
Yellow 2
Green 3
Cyan 4
Blue 5
Magenta 6
White 7
Gray 8

PREFIXES:

giga = G = 10^9
mega = M = 10^6
Kilo = k = 10^3
centi = c = 10^{-2}
milli = m = 10^{-3}
micro = μ = 10^{-6}
nano = n = 10^{-9}